Sustainable Development Goals Series

The **Sustainable Development Goals Series** is Springer Nature's inaugural cross-imprint book series that addresses and supports the United Nations' seventeen Sustainable Development Goals. The series fosters comprehensive research focused on these global targets and endeavours to address some of society's greatest grand challenges. The SDGs are inherently multidisciplinary, and they bring people working across different fields together and working towards a common goal. In this spirit, the Sustainable Development Goals series is the first at Springer Nature to publish books under both the Springer and Palgrave Macmillan imprints, bringing the strengths of our imprints together.

The Sustainable Development Goals Series is organized into eighteen subseries: one subseries based around each of the seventeen respective Sustainable Development Goals, and an eighteenth subseries, "Connecting the Goals", which serves as a home for volumes addressing multiple goals or studying the SDGs as a whole. Each subseries is guided by an expert Subseries Advisor with years or decades of experience studying and addressing core components of their respective Goal.

The SDG Series has a remit as broad as the SDGs themselves, and contributions are welcome from scientists, academics, policymakers, and researchers working in fields related to any of the seventeen goals. If you are interested in contributing a monograph or curated volume to the series, please contact the Publishers: Zachary Romano [Springer; zachary.romano@springer.com] and Rachael Ballard [Palgrave Macmillan; rachael.ballard@palgrave.com].

Saa Dittoh · Anna Bon
Hans Akkermans

Editors

Integrating Indigenous and Scientific Knowledge for Sustainable Food Systems in Africa

The Plug-In Principle

Editors
Saa Dittoh
University for Development Studies
Tamale, Ghana

Anna Bon
Vrije Universiteit Amsterdam
Amsterdam, Noord-Holland
The Netherlands

Hans Akkermans
AKMC Knowledge Management B.V.
Koedijk, Noord-Holland
The Netherlands

ISSN 2523-3084   ISSN 2523-3092   (electronic)
Sustainable Development Goals Series
ISBN 978-3-031-85511-5   ISBN 978-3-031-85512-2   (eBook)
https://doi.org/10.1007/978-3-031-85512-2

Prologue: Integrating Indigenous and Scientific Knowledge for Resilient, Sustainable, and Inclusive Food Systems

Air, water, and food are the three most basic requirements of all living things. Air, the most essential of all three, has been graciously provided free by nature (God) and water is relatively abundant because of nature's rainfall. Even though present-day activities worldwide are threatening both clean air and potable water, the greatest threat to humankind is food. Food has been produced over centuries in all parts of the globe in different ways within varying prevailing environmental conditions. However, overtime, various factors including technological advancements, population growth, industrialization, globalization, and others have reshaped and continuing to reshape how food in produced, distributed, and consumed. The food systems of different people in different parts of the globe have undergone and are undergoing changes, but unfortunately most of the changes have been in the wrong direction. According to von Braun et al. (2021), "the world's food system is in disarray, one in ten people is undernourished, one in four is overweight and more than one-third of the world's population cannot afford a healthy diet." This disturbing situation is traceable to the present wrong perception of the reason food is produced, and the focus on short run neoclassical "productivity-efficiency-profitability" paradigms, which often fail to address the critical issues of equity, resilience, and sustainability (Dittoh and Akuriba 2018). It is important to retrace our steps to the basic reason for food, which is, for our human bodies to be well nourished and healthy for production and reproduction over our lifetimes. Food systems practices currently being promoted cannot ensure this basic reason for food production. Food and nutrition insecurity in all its forms: protein-energy malnutrition (PEM), micronutrient malnutrition (MM), and overweight and obesity (OO) as well as non-communicable diseases are on the increase in all parts of the globe (FAO et al. 2022; WHO 2023). Another dimension of the prevailing industrial ("green revolution") food production system, which is based on monocultures of mainly wheat, rice, maize, and soybean, and massive use of heavy machinery, fertilizers, and agroechemicals, had been its devastating effect on biodiversity and agrobiodiversity, soil and water conservation, and food sovereignty, as well as its tendency to perpetuate land and water grabs, loss of jobs, and detrimental effects in rural economies and local communities.

Africa was food and nutrition secure in pre-colonial, colonial, and even post-colonial times until emphasis in agricultural production shifted from staple foodstuffs to export crops in the 1950s and 1960s. That trend has continued till date. The only change has been attempts to replace Africa's food production systems with the industrial ("green revolution) agricultural system. That has, fortunately, not been successful in most parts of Africa, but it has been a major contributor to Africa's deteriorating food and nutrition insecurity situation. Africa has wasted considerable large amounts of own and borrowed resources on the failed systems (Collins and Leppe 1979). The realization worldwide now is that indigenous knowledge (which is nature-based knowledge) has been ignored and can be a main reason for the failure of the introduced systems. There is need for introduced ("scientific") knowledge to "plug-in" into nature-based knowledge for the mutual benefit of both human beings and nature (the environment) in the short, medium, and long term.

This book presents theoretical arguments and practical approaches for the effective and sustainable integration of indigenous and scientific knowledge and technical know-how, toward the achievement of the SDGs, especially SDG2, in Africa and beyond. The chapters explore, through decades of field research and experiences in Burkina Faso, Ghana, Kenya, Mali, Nigeria, and South Africa, the various ways used to integrate traditional and local knowledge and practices, with scientific-based research; the challenges encountered, and how the "plug-in" concept can be used for effective integration of knowledge systems and effective development interventions. The various field examples show the value of experiential local knowledge and community-led innovations in improving food and nutrition security and promoting inclusive, resilient, and sustainable food systems.

The authors of the chapters recognize the limitations of smallholder farming systems with regards production resources including land and finance, markets, communication, technical knowledge, climate change, resource conflicts, and others, and the critical role appropriate and responsible scientific knowledge can play in ensuring resilient and sustainable food systems on the continent. What is not useful are technologies that disrupt existing livelihoods, compromise human and local values, destroy local cultures, exacerbate gender and generation inequalities, discriminate against minorities and underprivileged persons, degrade soil and landscapes, and destroy the natural environment.

The plug-in principle has been presented as an alternative to the dominant narratives of how to ensure appropriate and effective implementation of development interventions and the adoption of appropriate technologies to food production. It is a guiding theoretical and practical approach toward integration of knowledge systems. The principle is analyzed and critically assessed, without idealization or negative bias toward modernization. The book takes an African perspective, as most of the authors are African scholars with decades of experience in field-based research. The perspective of the African smallholder farmer, who is the main stakeholder and beneficiary and represents the largest producer of food on the continent, is central in all the studies. The plug-in principle starts from existing situations and tries to "better" the situations by plugging-in new ideas, techniques, and practices.

In academic research and teaching and learning, there is still a strong negative bias toward traditional knowledge systems. They are sometimes dismissed as non-scientific, under-developed, and primitive. Views are however changing, though very slowly, toward a fact that sustainable development is about planetary health, healthy people, and common prosperity, and should be in line with the spirit of the sustainable development goals of reducing inequalities, protecting the natural environment, reducing poverty, and enhancing food systems for all. The book aims at championing this realistic view of sustainable development and how to pursue it.

The book is structured into three parts but with a prologue and an epilogue. Part I, which has five chapters, covers the theoretical arguments of the concept and practical approaches based on field research and experiences. In Chap. 1, Saa presents the plug-in principle as a theory of integration as well as a practical approach to blend new ideas into existing wisdom. He, with Mojisola and Anna, in Chap. 2, describes how both Ghana and Nigeria have been groping in the dark with respect to their agricultural development paths because of the refusal to plug-in to the rich indigenous knowledge of Ghanaian and Nigerian smallholder farmers. In Chap. 3, Sandile and Sikhulumile from South Africa present results of a smallholder irrigation case study and conclude that government, private institutions, and NGOs should build on farmers' agencies to plug modern knowledge into the indigenous knowledge systems through appropriate research and innovations, rather than trying to replace indigenous systems with modern farming systems. Chapter 4 is another case study by Anna and a team of researchers from Mali and the Netherlands in which two parallel existing food eco-systems in rural Mali are described. One is the conventional development approach, aimed at increasing food production through the usual "green revolution" technologies, and the other is based on an alternative grassroots approach, aimed at protecting local autonomy and the right of seed producing farmers to develop and use their own knowledge and local practices to produce, harvest, trade, and consume. Chapter 5 is a gendered smallholder irrigation case study in northern Ghana by Mercy, Saa, and Lesley. It is an irrigation farming systems study and shows how different situations are for men and women even in farmer-led irrigation which is indigenous to both genders. The study also shows that women tend to be more conscious of indigenous irrigation practices including the use of organic inputs in production as compared to men but are constrained by several socioeconomic factors.

Part II of the book consists of four chapters. They describe several challenges to achieving sustainable food systems in Africa, and the role indigenous conflict resolution methods can play in reducing and/or removing them. Chapters 6, 7, and 8 are on farmer-herder conflicts in West and East Africa which are formidable threats to the achievement of the SDGs, while Chap. 9 focuses on the unfortunate neglect of traditional ecological knowledge (TEK) in South Africa. Chapter 6 by John and his colleagues at the UDS, Tamale, Ghana, discusses the efficacy of traditional dispute resolution mechanisms in facilitating peacebuilding in a case study on conflicts between herdsmen and farmers in the West Mamprusi Municipal of Ghana, while Moses in Chap. 7 dwells on the challenges faced by Maasai pastoralists in the Kajiado County of Kenya over several decades, and how the Kenyan government has refused to consider the people's rich knowledge to be part of the development agenda. That continues to affect the pastoralists negatively, and thus their livelihoods and food security. It is however obvious that a plug-in to their indigenous knowledge is all that is required for peace, progress, and better livelihoods for the Maasai. Madam Ida in Chap. 8 takes a legal look at the negative effects of dismissing the indigenous knowledge-based social justice system among the marginalized Maasai pastoralists of the Kajiado County. She argues that an acceptable and equitable justice system must be one that effectively integrates the indigenous and introduced formal systems, with the indigenous being the basis for integration. In Chap. 9, Chenai and Gladman bemoans the neglect of TEK in the Eastern Cape of South Africa despite the overwhelming evidence of its positive impact on natural resources management and ecosystems services enhancement in and outside Africa. They point out that the challenges faced are many and complex and have been because there has been a deliberate ineffective incorporation of the rural traditional systems into the South African nation-state. In other words, the TEK in the Eastern Cape had been sidelined and ignored by the state system, when there is need for it to be integrated into the formal system for effective management of the country's natural resources.

In Part III, the strength of indigenous knowledge and community-based practice is discussed in two case studies from Ghana. In Chap. 10, Saa, Paul, and Anna reflect on the need for students to learn and appreciate knowledge of the local environment and rural communities from the onset. Through the "third trimester field practical programme" (TTFPP), all students of the University for Development Studies (UDS), in Tamale, Ghana, acquire deep understanding of the local context, its barriers and limitations, and of community-based efforts and collective knowledge of how to leverage solutions to local problems related to agriculture, irrigation, soil and water conservation, health and nutrition, and social and economic issues. The chapter includes statements of experiences of three former students of UDS who went through the TTFPP and the impact the program had on them. In Chap. 11, Francis notes the importance of ICT4D for the promotion of sustainable food systems in rural communities that are often in hard-to-reach locations. He discusses some successes and failures in efforts to design and build novel, digital technologies to support local farmer communities in the Northern Region of Ghana. He argues that for success in ICT4D projects and interventions, there is need to follow "the ICT4D Plug-in Principle." Lastly in Chap. 12, Marie-Lou and Anna present a digital application that was used to study women entrepreneurs in their trade of non-timber forest products (NTFP) in Burkina Faso and concluded that it integrates digital and local knowledge in a practical way similar to what is suggested by the plug-in principle.

The cases presented in this book do not claim to provide all the solutions to the global or even African food and nutrition security challenges. It has only been an attempt to provide a major alternative view of effective integration of knowledge systems, with practical examples. There is considerable room for refinements of the ideas, and presentation of related viewpoints. The ICT chapters in the book suggest the important role ICT has to play in the agricultural and food sectors of African countries toward the achievement of the SDGs.

The "Plug-In Book," as we choose to call it, should be of great interest to a wide and diverse readership. It will be of immense benefit to teachers and students in varied disciplines in universities and colleges across the globe. Researchers, especially those involved in interdisciplinary and transdisciplinary research, will find the book handy and helpful. The simple language and careful explanations of the book will also make it appealing to policymakers, politicians, and general readers.

Vrije Universiteit Amsterdam   Anna Bon
Amsterdam, Noord-Holland, The Netherlands
University for Development Studies   Saa Dittoh
Tamale, Ghana
AKMC Knowledge Management B.V   Hans Akkermans
Koedijk, Noord-Holland, The Netherlands

References

Collins J, Leppe FM (1979) Whom does the World Bank serve? Econ Polit Wkly, 12 May
Dittoh S, Akuriba MA (2018) Africa's looming food and nutrition insecurity – a call for action. Ghana J Agric Econ Agribus 1(1):148–170
FAO, IFAD, UNICEF, WFP, WHO (2022) The state of food security and nutrition in the world (SOFI).
von Braun J, Afsana K, Fresco LO, Hassan M (2021) Food systems: seven priorities to end hunger and protect the planet. Nature 597(2):28–30
WHO (2023) Fact Sheets. Noncommunicable diseases (who.int)

Acknowledgments

The editors would like to thank the following organizations for their support: the Web alliance for Regreening in Africa W4RA (https://w4ra.org), and the EURIDICE Digital Europe Project (https://euridice.eu), which is co-funded as Project No. 101123121 by the European Union.

Contents

About the Editors

Saa Dittoh is an agricultural development and food systems economist at the University for Development Studies (UDS), Tamale, Ghana, and a distinguished fellow of the African Association of Agricultural Economists (AAAE). He has been a leading scholar in the political economy of African agricultural and food systems and has, for over four decades, been a strong advocate for the development of smallholder farmer production systems. He is a member of the Web Alliance for Regreening in Africa (https://w4ra.org). He holds an MPhil and PhD from the University of Ibadan, Ibadan, Nigeria.

Anna Bon is senior international project manager, researcher, and lecturer in the field of Digital Transformation and Global Development, at Amsterdam, The Netherlands. She directs the interdisciplinary research program, Web Alliance for Regreening in Africa (https://w4ra.org). She is coordinator of EURIDICE, a Digital Europe project developing educational programs in 15 countries (in Europe, Asia and Africa), centered on the Digital Society and its societal impacts (https://euridice.eu). She holds a PhD degree in Science and Technology Studies from Maastricht University, The Netherlands.

Hans Akkermans is an Emeritus Professor of Business Informatics at Amsterdam, The Netherlands, and is an adjunct professor at the University for Development Studies (UDS), Tamale, Ghana. He is a leading member of w4ra.org, the international collaborative research, consultancy, and education program Web Alliance for Regreening in Africa (https://w4ra.org). He is furthermore one of the initiators of the Vienna initiative on Digital Humanism (https://dighum.org). He holds a cum laude PhD degree in theoretical and nuclear physics from the University of Groningen, The Netherlands.

Contributors

Mercy Apuswin Abarike University for Development Studies, Tamale, Ghana

Hans Akkermans AKMC Knowledge Management B.V, Koedijk, Noord-Holland, The Netherlands

André Baart Vrije Universiteit Amsterdam, Amsterdam, The Netherlands

Anna Bon Vrije Universiteit Amsterdam, Amsterdam, Noord-Holland, The Netherlands

Mohamed Coulibaly Vrije Universiteit Amsterdam, Amsterdam, The Netherlands
Université des Sciences Juridiques et Politiques de Bamako, Bamako, Mali

Marie-Lou David Vrije Universiteit Amsterdam, Amsterdam, The Netherlands

Lucas de Lange Vrije Universiteit Amsterdam, Amsterdam, The Netherlands

Saa Dittoh University for Development Studies, Tamale, Ghana

Dzigbodi Adzo Doke University for Development Studies, Tamale, Ghana

Ida Gathoni Kenyatta University, Nairobi, Kenya

Lesley Hope University of Energy and Natural Resources, Sunyani, Ghana

Mojisola Olanike Kehinde Landmark University, Omu-Arun, Kwara State, Nigeria

Chenai Murata University of Vienna, Vienna, Austria

Moses Mwangi South Eastern Kenya University, Kitui, Kenya

Paul Kwame Nkegbe University for Development Studies, Tamale, Ghana

John Peter Okoro University for Development Studies, Tamale, Ghana

Sandile Phakathi Rhodes University, Makhanda, South Africa

Francis Saa-Dittoh University for Development Studies, Tamale, Ghana
Vrije Universiteit Amsterdam, Amsterdam, The Netherlands

Sikhulumile Sinyolo Human Sciences Research Council, Pretoria, South Africa

Gladman Thondhlana Rhodes University, Makhanda, South Africa

Wendelien Tuijp Vrije Universiteit Amsterdam, Amsterdam, The Netherlands

Abbreviations

ACB	African Centre for Biodiversity
ACDEP	Association of Church Development Projects (Ghana)
ADPs	Agricultural Development Projects (Nigeria)
ADR	Alternative Dispute Resolution
ADRM	Alternative Disputes Resolution Mechanism
AFD	Agence Francaise de Developpement
AfDB	African Development Bank
AOPP	Association des Organisations Professionnelles Paysannes
ASR	Automatic Speech Recognition
AU	African Union
AUC	African Union Commission
CA	Conservation Agriculture
CIDA	Canadian International Development Agency
CSA	Climate Smart Agriculture
DDA	District Departments of Agriculture (Ghana)
DNA	Direction Nationale de l'Agriculture (DNA)
DPs	Development Partners
ECOWAS	Economic Community of West African States
FAO	Food and Agriculture Organization (of the UN)
FGDs	Focus Group Discussions
FLI	Farmer-Led Irrigation
FLID	Farmer-Led Irrigation Development
FMNR	Farmer Managed Natural Regeneration
FNS	Food and Nutrition Security
GASIP	Ghana Agricultural Sector Investment Programme
GCAP	Ghana Commercial Agricultural Project
GEF	Global Environmental Facility
GIDA	Ghana Irrigation Development Authority
GR	Green Revolution
GTZ/GIZ	German Development Agency
HPG	Hybrid Peace Governance
HYVs	High Yielding (crop) Varieties
ICRISAT	International Crops Research Institute for the Semi-Arid Tropics
ICT4D	Information and Communication Technology for Development
ICTs	Information and Communication Technologies
IFAD	International Fund for Agricultural Development

IFPRI	International Food Policy Research Institute
IIED	International Institute for Environment and Development
IITA	International Institute of Tropical Agriculture
IK	Indigenous Knowledge
IKS	Indigenous Knowledge System
ILEIA	Information Centre for Low Eternal Input Agriculture
IRRI	International Rice Research Institute
ITT	International Technology Transfer
KIIs	Key Informant Interviews
LEISA	Low External Input and Sustainable Agriculture
MCE	Municipal Chief Executive (Ghana)
MDGs	Millennium Development Goals
MK	Modern Knowledge
MM	Micronutrient Malnutrition
MMS	Massively Multilingual Speech
MUSEC	Municipal Security Council (Ghana)
NGLWG	Northern Ghana LEISA Working Group
NGOs	Non-Governmental Organizations
NRGP	Northern Rural Growth Project
NTFPs	Non-Timber Forest Products
OO	Obesity and Overweight
OP	Organisations Professionnelles
PEM	Protein Energy Malnutrition
PLEC	People, Land and Environmental Change
PNDC	Provisional National Defence Council (Ghana)
RBDAs	River Basin Development Authorities (Nigeria),
RBRDAs	River Basin and Rural Development Authorities (Nigeria)
RDCs	Regional Development Corporations (Ghana)
RDP	Reconstruction and Development Programme (South Africa)
RSA	Republic of South Africa
RTIP	Rural Technology Improvement Project (Ghana)
SADC	Southern African Development Community
SAI	Sustainable Agricultural Intensification
SAPIP	Savannah Zone Agricultural Productivity Improvement Project
SDGs	Sustainable Development Goals
SLM	Sustainable Land Management
SPG	Système Participative de Garantie
SSA	Sub-Saharan Africa
TDRMs	Traditional Dispute Resolution Mechanisms
TEK	Traditional Ecological Knowledge
TNCs	Transnational Corporations
TTFPP	Third Trimester Field Practical Programme (Ghana)
TTS	Text-To-Speech
UDS	University for Development Studies (Ghana)
UNDP	United Nations Development Programme
UNIMAS	University of Malaysia
UNU	United Nations University (Tokyo, Japan)

URADEP	Upper Regional Agricultural Development Programme (Ghana)
USAID	United States Agency for International Development
VORADEP	Volta Regional Agricultural Development Programme (Ghana)
VU	Vrije Universiteit Amsterdam (The Netherlands)
W4RA	Web Alliance for Regreening in Africa
WFP	World Food Programme
WMMA	West Mamprusi Municipal Assembly (Ghana)
WUA	Water User Associations

List of Figures

List of Tables

Plug-In Principle: Theory and Practical Approaches to Integration of Knowledge Systems

The Plug-In Principle: A Theory for Effective Integration of Knowledge Systems and Development Interventions

Saa Dittoh

Abstract

Africa does not lack resources to propel sustainable food and nutrition security for all. The problem has been systematic attempts at replacing supposedly inferior knowledge systems with those thought to be superior. That had been the main cause of lack of sustainability and resilience of development efforts, especially in developing countries. No knowledge system can claim superiority over another. Effective integration of knowledge systems is what will result in the achievement of the SDGs, especially SDG2. The Plug-In Principle describes a methodology for the effective integration of indigenous and "modern" knowledge systems to ensure sustainable outcomes of development interventions. The two-way plug-in, in the right order, ensures increased productivity, resilience and sustainability of agricultural and other systems.

Keywords

Food and nutrition security · Knowledge systems · Indigenous knowledge · Replacement mentality · Effective integration · Plug-In

S. Dittoh (✉)
University for Development Studies, Tamale, Ghana

1.1 Knowledge Systems and the Importance of Integration

Food and nutrition security has been and will continue to be a basic requirement in all countries of the world. Food and nutrition security is necessary in the short, medium and long term. It is not a passing phenomenon; thus, sustainable food and nutrition security is what must be pursued. Unfortunately, in almost all parts of the world, emphasis has been on short term achievements of food and nutrition security. That and other shortcomings of the world development systems gave rise to the need for the millennium development goals (MDGs) and then the sustainable development goals (SDGs) (UNDP and World Bank 2015; Griggs 2015). Developing countries, in particular, have been facing a clash of knowledge systems with respect to food production, distribution and consumption. Should production be small scale or large scale? Should distribution be localized or globalized? Should rice and maize replace millet and sorghum as staple crops? Should roots and tubers be abandoned because they are less responsive to modern production inputs? There have been these and many similar unanswered questions. All the questions need answers and/or clarifications if sustainable food and nutrition security is to be pursued in all countries of the world. The clash of knowledge systems must be resolved. Sustainability, though

S. Dittoh et al. (eds.), *Integrating Indigenous and Scientific Knowledge for Sustainable Food Systems in Africa*, Sustainable Development Goals Series, https://doi.org/10.1007/978-3-031-85512-2_1

difficult to define adequately, because it is often viewed differently by different persons and under different situations, is valued by all persons. The common view, that it involves meeting the needs of the present without comprising the needs of future generations, is often agreeable to most people even though the actions of many people may be contrary. This means in theory, people of whatever knowledge system or systems they profess, seek sustainability and thus the effective integration of knowledge systems is important. It is the interaction of different knowledge systems that generate new knowledge. According to Buzinde et al. (2020), co-produced solutions (from different knowledge systems) are essential for sustainability outcomes. Achievement of the SDGs, and especially SDG 2, depends critically on the integration of several food and other knowledge systems. Achievement of sustainable and resilient food systems at all levels require integration of knowledge systems at all levels of the food chain: production, post-harvest handling, processing, marketing, consumption and post-consumption. This chapter looks broadly at the integration of knowledge systems and why developing countries, in particular, should avoid replacement, and pursue effective integration, of knowledge systems to ensure sustainability of systems generally and food systems in particular. The neglect of indigenous knowledge systems tends to engender unsustainable outcomes of development interventions.

There have been arguments against the usefulness of indigenous knowledge (IK). Some of the arguments are that "science (often termed Western science) radically differs from all traditional knowledge systems" (Agrawal 2009), and IK and science have "different epistemologies and should not be blended" (Moller et al. 2009); implying that integration of IK and science (Western science) cannot result in any positive outcome. There have however been equally strong arguments for positive outcomes of the integration of IK and science (Veitayaki 1997; Altieri 2004; Berkes 2009; Maclean and Cullen 2009; Athayde et al. 2017). Ludwig (2016), for example, sees knowledge integration to be "often epistemically productive and socially desirable".

Manuel-Navarrete et al. (2021), however, argue that to effectively integrate IK and science, there is need "to end the monopoly of science" while Latulippe and Klenk (2020) point out that there is a need to "de-center Western science and institutions as primary sites of knowledge production and leadership". These arguments are not about the relative superiority of knowledge systems but the importance of their effective integration or what Manuel-Navarrete et al. (2021) refer to as "horizontal co-production and decolonization". At the practical level, Dittoh (2003) has given examples of how the integration of IK and science has led to better management of natural resources, improved and sustainable food production practices, and better food preparation and consumption practices in several parts of northern Ghana. There are several other practical examples in the food and agriculture sector that give evidence of positive outcomes of the integration of IK and science (Williams et al. 2020; Bwambale 2023). According to Bwambale (2023), "the recognition of limitations and overlapping strengths between science and indigenous knowledge widely prompts arguments to integrate science and indigenous knowledge". The current rise of agroecological practices worldwide as resilient and sustainable agricultural practices is good evidence of the importance of the integration of IK and science (Altieri 1995, 2004; Pretty et al. 2011; IPES-Food 2018; Chappell et al. 2018).

Knowledge systems have been increasing and improving over centuries and a trace of the trend of some knowledge systems over time can support arguments against over prioritization of some knowledge systems. History and religion (Judaism, Christianity and Islam) has it that civilization started in Mesopotamia at the confluence of the Rivers Tigris and Euphrates (present day Iraq). That means the first knowledge system might have started there (in the garden of Eden).[1] Since then, the world has witnessed thousands, probably millions, of different knowledge sys-

---

[1] This does not in any way purport to claim that the Garden of Eden is the origin of knowledge. What we do know, and many people accept, is that some knowledge started there.

tems and every one of them started somewhere. That means every knowledge system has been "indigenous" to the people who started it. However, every knowledge system has changed or advanced and had elements that "died" over time. The question that should be asked is what engenders or causes or engineers the start of a knowledge system? The simple answer is a desire to meet a need and/or solve a problem. According to religion, God, at the beginning of man's creation, had a desire to create man and make him have dominion over all things and to undertake agriculture, so He engineered the "garden of Eden" knowledge system, an irrigation knowledge system. That irrigation knowledge system has continued till today with numerous modifications. It is interesting that the Rivers Tigris and Euphrates and their tributaries are still important sources of water for irrigation in Iraq, Turkey, Syria and other places till date. The irrigation knowledge system continued to be very relevant in early times. The River Nile was and continues to be a major source of water for agriculture in several African countries. The people of old in the Middle East always found sanctuary in Egypt because of the agricultural produce of the River Nile.

Other knowledge systems, such as gathering and hunting, also emerged in areas where crop and animal production knowledge systems were not yet known. The hunting and gathering arose out of the need for the people to survive. Those systems also exist till date, though very marginally, because improved systems that can meet the same objectives have emerged. So, in most societies the hunting and gathering knowledge systems have either been completely replaced or new systems have been integrated into them. The hunting knowledge system has turned into a game in several societies, but gathering still exists in areas where some crops/trees and animals have not been domesticated. Products from shea (*Vitellaria paradoxa*), dawadawa (*Parkia biglobosa*), baobab (*Adansonia digitata*) and other semi-wild trees and animals as well as some indigenous vegetables, for example, are still being gathered in Africa because they have not been effectively domesticated. The defense

knowledge system using available materials evolved with the hunting knowledge system since people had to defend themselves against wild animals. The simple defense knowledge system has now been developed into a very sophisticated system capable of destroying the whole world, but it is good to recognize its simple origin.

The present-day medical knowledge system also has its origin in the use of locally available medicinal plants to treat diseases. Herbal medicine continues to be important in several societies despite the great advancement in medical practice. The recent COVID 19 pandemic has regenerated great interest in medicinal plants in Africa and other places. The present sophisticated engineering knowledge systems also grew out of the knowledge of using available materials (of wood and iron) to produce tools to lessen human effort used in the production of food and other necessities.

The above discussion is to point out that every knowledge system has an indigenous origin. Thus "people's science" or "ethno-science", that is indigenous knowledge, is the origin of every knowledge system. That is why indigenous knowledge systems do not die. They are "accepted lived experiences". They have proved to be resilient. Their "souls" are imbedded in communities in the modern systems. The "souls" of irrigation in the garden of Eden, of gathering and hunting, of use of bows and arrows for defense, herbs for curing diseases, medieval wood and iron technology for food production etc. are still existing despite massive "modernization". It has been argued that the current sophisticated ICT system with AI originated from indigenous virtual forms of sending messages between households and communities by high-pitched voices and/or loud sounding talking drums that gave coded information that were understood by those who had the knowledge and/or skills (Dittoh 2021). Also, according to Mapara (2009) knowledge forms (or systems) in Zimbabwe have failed to die despite racial and colonial onslaught over time. Indigenous knowledge systems are only resurfaced and transformed in the face of modern contradictions.

To simplify the understanding of the integration of knowledge systems, they (that is, knowledge systems) may be broadly divided into indigenous and "modern" knowledge systems (IK and MK). All modern knowledge (MK) systems are, however, derived from, or are modifications, of indigenous (IK) systems. Indigenous knowledge systems are bodies of knowledge of indigenous people of particular geographical areas that have survived over a long period of time. They are forms of knowledge and knowledge practices that have originated locally and naturally (Altieri 1995). As explained earlier, every knowledge system has been indigenous to the people who started it. However, knowledge systems have changed or advanced over time.

Dittoh (1981) described how agriculture in England (from late 1700s), the US (from late 1800s), Japan, the then USSR and China all evolved from the people (i.e. IK) and how the successes of the green revolution in the Indian sub-continent was short-lived (mainly because it did not evolve from the people). Lester Brown of the American Overseas Development Council who was deeply involved in the Asian green revolution programme had to agree that the green revolution did not represent a solution to the food problem; it was a means of buying time (Brown 1975). Several researchers led by Robert Chambers have since the July 1987 "Farmer First" workshop at the Institute of Development Studies (IDS), Sussex, United Kingdom (Chambers et al. 1989), advocated strong integration of farmers' knowledge (IK) and MK for sustainable agricultural development. Later the Information Centre for Low Eternal Input Agriculture (ILEIA), established in 1984 in the Netherlands, spearheaded the low external input and sustainable agriculture (LEISA) concept which largely promoted IK and its integration with other knowledge systems. ILEIA posited that "Green Revolution technology was neither sustainable nor feasible for many small-scale farmers around the world". No wonder that in Africa, "despite generations of western influence, the decisions about agriculture, health and nature

management are still heavily based on the concepts of African traditions" (Millar and Dittoh 2004). The basis for work done by researchers at the Vrije Universiteit, Amsterdam and four African teams in North, West, East and Southern Africa between 2013 and 2016 was that "there are local community innovations succeeding where formal research recommendations have often failed" (Mudhura et al. 2016). Current research continues to point out that lack of integration of IK and MK makes education and research "too academic and distant from the developmental challenges of African local communities" (Kaya and Seleti 2013; Offei et al. 2010; Teeken et al. 2012; Mokuwa et al. 2013). Teeken et al. (2012) showed that ecological and socioeconomic factors as well as cultural norms and values, shape the use and development of local rice technologies, while Mokuwa et al. (2013) showed that West Africa farmers' rice varieties are robust and tolerant of sub-optimal conditions. Navarro et al. (1994) have also argued that hermetic storage[2] is not a new concept and has been recorded from prehistoric times when grain was stored underground in pits. They argue that hermetic bags are an adaptation of the traditional and other forms of hermetic storage systems such as underground pits, clay pots, jerrycans, silos, or drums.

The argument is that sustainable development processes must be based on effective integration of knowledge systems. If MK systems grew out from their indigenous origins because of the need to solve specific problems of societies, it makes sense that the best way for two knowledge systems from different environments to be useful to societies is for their effective integration. The integration should, however, be a two-way process in the right order, to ensure resilience, sustainability and long-term productivity.

---

[2]Hermetic storage is being promoted for small farmers in Africa and elsewhere mainly because it is a chemical-free, safe and simple means of storage of all types of dry grains (cereals and legumes).

1.2   The Replacement Mentality of Modernization

Despite the scientific findings of the importance of IK in sustainable agricultural production, "development models" being pursued in Africa and elsewhere currently, by Governments, development partners (DPs) and several NGOs, typically try to ignore IK. Development interventions in Africa continue to insist on REPLACING IK systems with "modern" systems (generally termed technology transfer). Several African countries have over several decades drawn programmes and instituted projects to "modernize" their agricultural and other sectors. World Bank assisted Agricultural Development Projects (ADPs) in Nigeria and Nigeria's River Basin Development Authorities (RBDAs), Ghana's Regional Development Corporations (RDCs), as well as several Africa Development Bank (AfDB) and Alliance for Green Revolution in Africa (AGRA) assisted agricultural projects in all parts of Africa are examples of such.[3] What most of the programmes and projects have attempted for several decades is to "replace" the practices of the people, and for decades there have been consistent and persistent failures. Several reports and research papers have continued to bemoan the non-adoption and low adoption of modern technologies and have put the blame on practitioners, mainly farmers (Feder et al. 1981; Acheampong et al. 2018; Mapemba et al. 2019; Kebebe 2019; Balehegna et al. 2020; Kazembe 2021). This process of blaming people (especially farmers) for resisting unproven and indeed destructive modern technologies has been very unfair. It has been based on a strange mentality of "the know-all" teaching the "know-nothing" and the latter should have no say in what the former believe is good for them. As stated by Horowitz (1990), "development interventions (in Africa) designed to enhance the habitat and improve the income and productivity of small-scale rural households so often have exactly contrary outcomes". It is the reason the people do not willingly adopt many of the so-called modern technologies.

Most of the attempts at modernizing Africa's agriculture are watered-down versions of the barely successful Asian green revolution;[4] and so far, there are hardly any evidence of significant achievements (Dawson 2016; Dittoh and Akuriba 2018; Bassermann and Urhahn 2020; Wise 2020). In Africa's agricultural sector, attempts have been made to REPLACE (not improve) the hoe and cutlass, seeds, breeds, indigenous methods of soil amendments, farmer-led irrigation systems, indigenous soil and water conservation systems, land tenure systems, cropping systems, livestock production systems, indigenous agro-forestry systems etc. and hardly any of the replacement attempts can claim any sustainable success. According to FAO (2016), African farm systems are the least mechanized across all continents. Only 1% of all households across six countries (Ethiopia, Malawi, Niger, Nigeria, Tanzania, and Uganda) own or hire tractors (Sheahan and Barrett 2017). It is so despite the fact that there have been substantial state-led efforts in Africa to promote mechanization during the 1960s and 1970s, often involving public machinery imports (Pingali 2007). Most of those efforts, however, failed due to governance challenges according to Pingali (2007). They actually failed because of the "replacement mentality", and the situation is unlikely to change unless there is a change in that mindset. In recent years, there has been talk of use of "Uber-like" mobile mechanization systems (Daum et al. 2021) but that is still begging the critical problem facing agricultural mechanization in Africa. Daum and Birner (2020) have correctly pointed out that there is need for research areas that will help to find mechanization solutions in Africa that are economically, socially, and environmentally sustainable. It is interesting that the Food and Agriculture Organization (FAO) and the African Union Commission (AUC) said in 2018 that "doubling agricultural productivity and eliminating hunger

---

[3] See Chap. 2 for discussion on some of these agricultural development projects.

[4] B. H. Farmer has edited a book, "Green Revolution?" (Farmer 1980) which has discussed extensively "the large question-mark" of the Asian Green Revolution.

and malnutrition in Africa by 2025 will be no more than a mirage unless mechanization is accorded utmost importance" (FAO-AUC 2018) and yet there is hardly any direction to mechanization development in Africa in 2025. There is a need to start from the basics and remove the replacement mentality. Similar situations can be seen in the seeds and breeds, extension, irrigation and several other sectors and sub-sectors in the agricultural, industrial, health and even food consumption sectors across Africa. Development of every type must be situated within the culture of the people for it to be relevant (Apusigah 2011). Any development initiative which does not integrate well with the culture of people will result in mal-development, and culture is derived largely from IK.

The replacement mentality in agriculture and possibly other sectors have arisen out of the erroneous belief that "modern" systems are superior and indigenous (or local) systems are inferior and that supposedly superior knowledge systems should replace supposedly inferior ones. That has not worked in many situations. That mentality goes far down to culture and social values where prevailing Western social values are not only considered to be superior to all others but are superior to past Western social values. The mentality continues to stifle development in many societies, including the Western societies themselves, because the supposedly inferior societies try to imitate the "superior" societies by abandoning useful and necessary societal values. Societies that have sought development based on local values have been successful most of the time. Maclean and Cullen (2009) point out that "working with indigenous Australians as research partners (rather than end-users of research) is one pathway for the (effective) integration of indigenous and scientific knowledge". Benneh (2011) has also pointed out that "Japanese agricultural transformation was based on improvement on their native technology (the Meji technology)". That can be said of several other successful development models across the globe.

The continued calls for African and other farmers to adopt modern technology are premised on claims that in the agricultural sector, "modern" systems are more productive, efficient and profitable. Several researchers have, however, questioned those claims. As far back as 1975, Uma Lele undertook an in-depth review of 13 donor-funded projects in seven African countries, and concluded that, "except for a few well-conceived export crop schemes (such as smallholder tea and coffee projects in Kenya), integrated rural development projects, as practiced by donors at the time in Africa, had limited impacts and even bigger questions about their sustainability" (Lele 1975). There has been growing agreement that conventional Green Revolution agricultural intensification strategies based on high yielding crop varieties and inorganic fertilizers cannot achieve the desired outcomes of increasing agricultural productivity sustainably, providing adequate nutrition and avoiding environmental destruction (Farmer 1980; Pingali 2012; Kim et al. 2019). Also crop yield responses to fertilizers have been very low (Dittoh et al. 2012; Kassie et al. 2013; Sheahan and Barrett 2017). One recommendation has been a turn to sustainable agricultural intensification (SAI) (Tilman et al. 2002; Petersen and Snapp 2015; Kim et al. 2019). The goal of SAI is "producing food from the same area of land while reducing the environmental impacts" (Godfray et al. 2010). However, obtaining the right outcomes in food-related issues is complex. It will require drastic changes in the food production, distribution and consumption systems especially in this era of climate change. So many other recommendations since the realization of the destructive effects of climate change have been climate smart agriculture (CSA), conservation agriculture (CA), regenerative agriculture, agroecology and several others. Most of these however have their origin in IK. Agroecological systems, for example, are deeply rooted in the ecological rationale of traditional small-scale agriculture (Toledo 1990; Altieri 2004). There have also been arguments that agroecological systems have fed much of the world population for centuries and continue to feed people in many parts of the planet, especially in developing countries, and hold many of the potential answers to the production and natural resource conservation challenges affecting today's rural landscapes (Koohafkan and Altieri 2010).

Hazell (2013) also argues that "small farmers are the most economically efficient producers", while others hold the view that family farming which main features consists of the preservation of traditional knowledge, sustainable management of natural resources, the empowerment of women and an economic model based on community and solidarity, is the key element in fighting hunger consistently and effectively (Caritas Europa 2014). Dittoh and Akuriba (2018) have also stated that the overemphasis on short term productivity (yields) and profitability goals has been the bane of African agriculture while Muya (2007) has argued that many modern systems in Africa lack indigenous content and thus "tend to be inadequately sensitive to the developmental challenges of local communities and country". This is not an argument against the usefulness of MK but the importance of integration of IK and MK. The argument is for modern technology to grow out of IK within the environments of the IK, and then probably take the lead since more science would have been imbedded in the MK. Most MK systems do not grow out of IK systems prevailing within their specific environments. They are often developed in different environments and then "tested" for technical suitability in other environments. That methodology cannot result in acceptance and sustainability of the developed technology. Sustainability has economic, environmental as well as socio-cultural dimensions and all the dimensions must be important considerations of a sustainable technology.

1.3 Effective Integration of Knowledge Systems: The Plug-In Principle

Development interventions have taken place in various developing, including African countries, over the years but the results have largely been disappointing. Apart from the Uma Lele study referred to above (Lele 1975), several development interventions especially in the agriculture sectors in Africa have fallen far short of the expected results. In many situations there is hardly a trace of the projects a few years after they end (ACB 2016a, b; Dawson 2016; Dittoh and Akuriba 2018; Bassermann and Urhahn 2020; Wise 2020). It has been mainly because interventionists (development workers) have been trying to FORCE change. The system of replacement has been to engender forced change and that is not the way to go. Change that is sustainable can only be NEGOTIATED between equal and mutually agreeable partners; it cannot be forced. That means development and dissemination of appropriate technologies must be by a "bettering process" and not a replacement or transfer or change process. Sustainable bettering process is what is illustrated by the Plug-In Principle (Fig. 1.1). It illustrates a principle for effective integration of IK and MK to ensure sustainability. The Plug-In Principle is based on the premise that interventions though important cannot replace what already exists. They can only improve upon or "better" them through co-creation and/or co-innovation processes. Thus, interventionists can only be "bettering agents" and not change agents. The principle also hinges on the belief that sustainable development is basically a continuous marriage of ideas or integration of knowledge systems, and every stakeholder has an important role to play. It is also a two-way process where MK grows out of IK and should take over as the lead, for a reverse "plug-in" by IK (Fig. 1.5).

Figure 1.1 depicts an IK pipe (A) which is wider than an MK pipe (B). The B pipe represents the "desired mindset" of an interventionist (development agent or worker). It illustrates that effective integration of knowledge systems must start with a "correct" mindset. The mindset of an interventionist at the beginning should be that what is already happening (IK) is more important or more relevant than what is being plugged-in (MK). Any other mindset will greatly compromise achievements. Such a mindset will prevent top-down intervention tendencies. With such a mindset, the interventionist will recognize the need to study, understand and appreciate what is in A (IK), that is, what the people know or do not

Fig. 1.1 Effective integration process—the plug-in principle

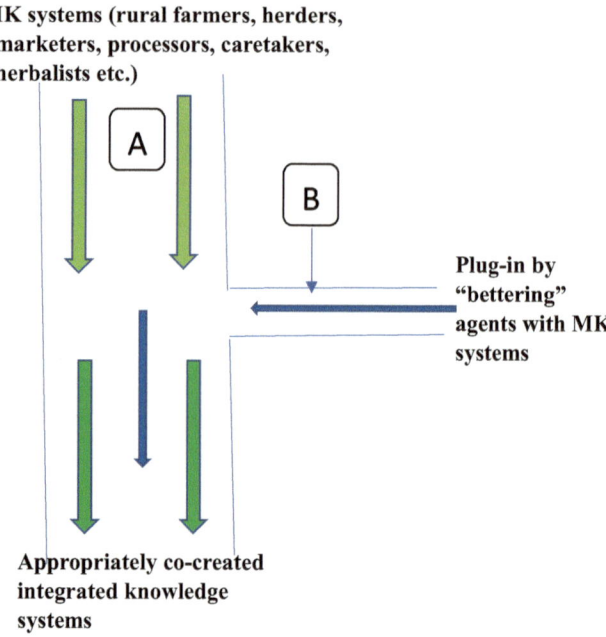

Appropriately co-created integrated knowledge systems

know and their mindset. It is this initial understanding and appreciation of IK and the people (usually called beneficiaries) that will help an appropriate design of a plug-in (MK). Interventionists must realize that the people or community will exist irrespective of what they do or do not do. It means the survival of the people or community is not so much dependent on any intervention. The intervention only "betters" the people's situation. It does not "change" their situation. Thus, development workers, researchers, donors and even governments should consider themselves as "bettering agents". Interventionists (i.e. the "bettering agents") should even see themselves as "intruders" and should bear the onus of success or failure of the interventions. Often, when interventions fail, the "expected beneficiaries" are blamed for non-adoption or non-responsiveness. That narrative must change. Whether an intervention succeeds or fails depends more on what the interventionist does or does not do than on the action or inaction of those already in a process. Sometimes incentives must be given by interventionists to encourage farmers to adopt innovations. However, when a plug-in with incentives must be continuous, then there should be something wrong with it.

Researchers and extension workers also hardly tell farmers the whole truth about "improved" technologies. Farmers, for example, are not usually told the simple truth that produce (grain) from hybrid seeds cannot germinate (they are "dead") or use of fertilizer over time on the same plot can lead to acidity or alkalinity. When farmers find out the truth later, they lose trust in extension messages. Appropriate "plug-in" ensures that the whole truth about a new technology is shared with farmers leading to the building of trust between interventionists and the people. Trust is a critical ingredient in all types of relationships and collaborations.

A measure of sustainability will be the extent to which the "plug-in" can be narrowed without compromising results. The plug-in should continue to get narrower over the years (Fig. 1.2) and ultimately become unnecessary. Successful intervention (plug-in) is obtained if "plug-in" can be narrowed without compromising results. Short-term interventions hardly achieve anything sustainable because they end when it is not yet time to narrow the plug-in. Some end when interventionists are even still trying to understand the IK systems (that is at the A stage).

Fig. 1.2 Narrowing of plug-in and positive impact—measure of sustainability of process

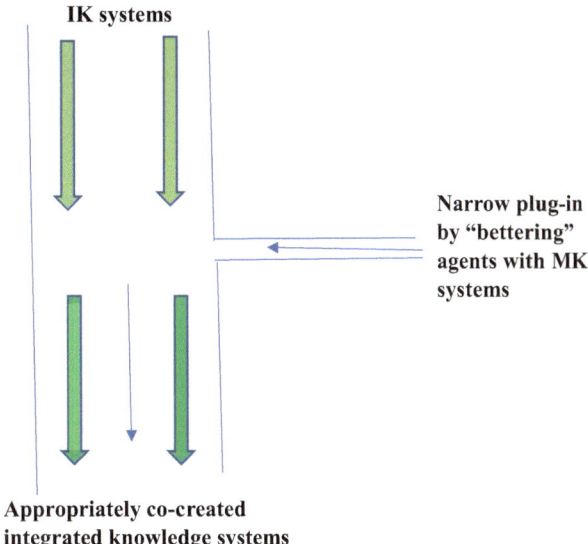

IK systems

Narrow plug-in by "bettering" agents with MK systems

Appropriately co-created integrated knowledge systems

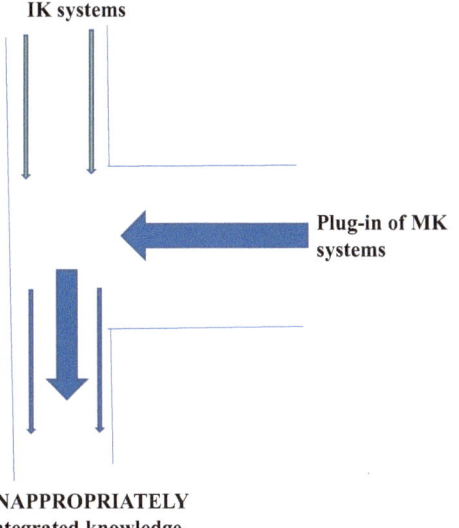

IK systems

Plug-in of MK systems

INAPPROPRIATELY integrated knowledge

Fig. 1.3 Wrong plug-in—top-down mindset from beginning

Figures 1.3 and 1.4 depict wrong plug-ins, that is, ineffective knowledge integration processes. Figure 1.3 depicts a "top-down" mindset where the interventionist's aim is to "civilize the uncivilized" or the more acceptable term "technology transfer". The concept of technology transfer is a knowledge system replacement, and no society will accept such unless it is forced. As explained earlier, forced interventions are not sustainable.

In Fig. 1.4, the pipe of the MK system is narrower suggesting that the "bettering" agent has a correct mindset, but the actions go contrary to the knowledge and aspirations of the people. May be the interventions go against their traditions, beliefs and/or values. That will also be resisted if not rejected outright.

The process should not, however, end with Plug-In by interventionists (MK). After the co-creation of the "appropriately integrated knowledge system" and the narrowing of the Plug-In (Fig. 1.2), the MK can take lead in the development process since the people would have had a good understanding of the process, and the intervention (MK knowledge system) would have been well integrated with the people's culture. The process would have provided joint understanding of what is possible and works and what does not. MK would now be in a good position to accelerate the development process, but with continuous "plug-in" by IK as shown in Fig. 1.5. The lead by MK at this stage is necessary because the "appropriately integrated system" would now be able to effectively enter the global system. In the area of agriculture, for example, the links to local and international markets, the development of commodity value chains, increases in produc-

Fig. 1.4 Wrong-
plug-in—intervening
against the values and
practices of the people
(Ignoring the importance
of IK)

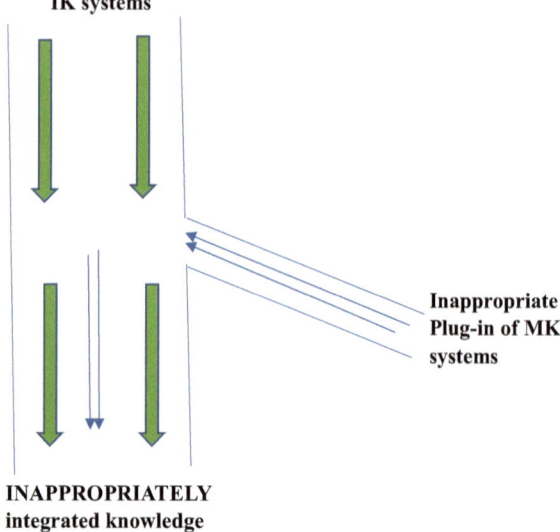

Fig. 1.5 Reverse
continuous Plug-In by
IK (for resilient,
sustainable and
productive systems)

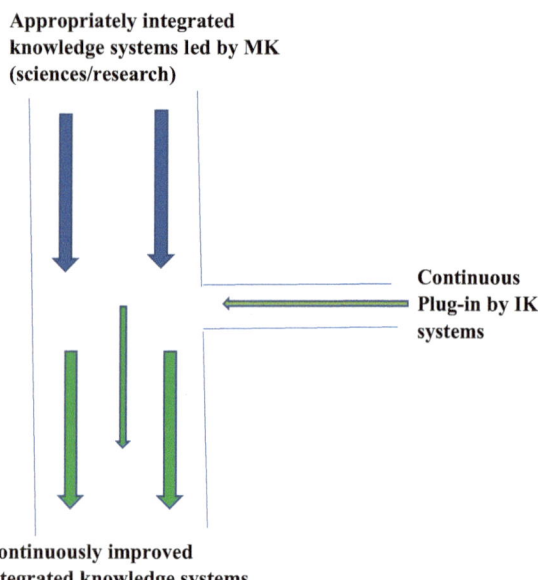

tivity and innovations (by the process of co-creation) will at this stage be more effectively driven by MK research (with IK plug-in).

1.4   Conclusion

The persistent African food and nutrition insecurity situation has been difficult to understand. The continent has far more than is required to produce more than enough food for its present and future populations. Africa's uncultivated arable land is about 60% of the world total. Its tropical climate is also probably the best for the production of a wide range of foods. Several parts of Africa can produce two annual crops a year without irrigation. It has agro-ecologies that range from mangrove swamps to semi-arid and arid landscapes and thus almost all types of crops and livestock can thrive. Africa's rivers, the Nile,

Congo, Limpopo, Niger, Volta, Gambia, Senegal and others with their numerous tributaries traverse the continent and groundwater is abundant. There is therefore very little reason for Africa's food and nutrition insecurity. It is instructive that except in conflict areas and severe drought years, most African countries did not experience serious food problems in the 1940s and 1950s. Even presently countries that produce cassava, yam, plantain, maize, sorghum, millet, the pulses, several horticultural crops are self-sufficient in them. The main foodstuffs Africa uses billions of dollars to import are rice, wheat, vegetable oils, meats and sugar. Except for wheat, all other crops can be produced in sufficient quantities even with current consequences of climate change and variability.

The problem therefore seems to lie in the unfortunate replacement of what nature bestowed on Africa (and other continents) with what is considered "modern". Scientific knowledge is indispensable, but it should not try to replace "nature". It should be used to "better" nature and that is the thrust of the SDGs. Johnston and Kilby (1975) argued strongly for "progressive modernization of millions of small-scale farmers" and stated that the starting point of the development process is "the traditional farmer in a traditional economy", while Hazell (2013) pointed out that "the most obvious way to make agricultural growth pro-poor is to engage with huge numbers of small farms". Those and similar views had been ignored in favour of technology transfer. The technology transfer model of agricultural development has, however, consistently failed because it is based on replacement (change) mentality. What is relevant is effective integration of indigenous and modern knowledge systems—a co-engaged betterment mentality. The Plug-In Principle explains the process of co-engagement and effective "bettering" of nature or indigenous knowledge. Africa and the world will be a better place if the emphasis is bettering nature in the quest to better humanity.

References

ACB (2016a) Green revolution dead-end in Malawi: two case studies—AGRA's Pigeon Pea Project and Malawi's Agro-Dealer Strengthening Programme (MASP). Available at http://acbio.org.za/wp-content/uploads/2016/06/Chinsinga-Report-ACBio-2016-06.pdf

ACB (African Centre for Biodiversity) (2016b) Farm Input Subsidy Programmes (FISPs): a benefit for, or the betrayal of, SADC's small-scale farmers? Available at http://www.db.zs-intern.de/uploads/1468387794-Input-Subsidies-Report-ACBio.pdf

Acheampong P, Amengor NE, Wiredu AN, Frimpong BN (2018) Does awareness influence adoption of agricultural technologies? The case of improved sweet potato varieties in Ghana. Ghana J Agric Econ Agribus (GJAEAB) 2(1):47–60

Agrawal A (2009) Why "indigenous" knowledge? J R Soc N Z 39:157–158

Altieri MA (1995) Agroecology: the science of sustainable agriculture. Westeview Press, Boulder, 433 pp

Altieri MA (2004) Linking ecologists and traditional farmers in the search for sustainable agriculture. Front Ecol Environ 2:35–42

Apusigah AA (2011) Indigenous knowledge, cultural values, and sustainable development in Africa. In: 2nd Annual Ibadan sustainable development summit, Nigeria, pp 1–15

Athayde S, Silva-Lugo J, Schmink M, Kaiabi A, Heckenberger M (2017) Reconnecting art and science for sustainability: learning from indigenous knowledge through participatory action-research in the Amazon. Ecol Soc 22(2):36. https://doi.org/10.5751/ES-09323-220236

Balehegna M, Duncan A, Tolera A, Ayantunde AA, Issa S, Karimou M, Zampaligré N, André K, Gnanda I, Varijakshapanicker P, Kebreab E, Dubeux J, Boote K, Minta M, Feyissa F, Adesogan AT (2020) Improving adoption of technologies and interventions for increasing supply of quality livestock feed in low- and middle-income countries. Glob Food Sec 26:1–11, 100372

Bassermann L, Urhahn J (2020) False promises: the alliance for green revolution in Africa. Biodiversity and Biosafety Association of Kenya (BIBA), Brot für die Welt (Germany), FIAN Germany, and others

Benneh G (2011) Technology should seek tradition – studies on traditional land tenure and smallholder farming systems in Ghana. Ghana Universities Press, Accra, 267 pp

Berkes F (2009) Indigenous ways of knowing and the study of environmental change. J R Soc N Z 39:151–156

Brown L. 1975. World population and food supplies. Agricultural initiatives in the third world. The Agribusiness Council. Lexington Books

Buzinde CN, Manuel-Navarrete D, Swanson T (2020) Coproducing sustainable solutions in indigenous communities through scientific tourism. J Sustain Tour 28(9):1255–1271. https://doi.org/10.1080/09669582.2020.1732993

Bwambale B (2023) Chapter 9: Integrating indigenous knowledge with science to suitably tackle disasters due to climate and environmental change: an overview of the progress and way forward. In: Pal I, Shaw R (eds) Multi-hazard vulnerability and resilience building: cross cutting issues. Elsevier, pp 127–143. https://doi.org/10.1016/B978-0-323-95682-6.00008-5

Caritas Europa (2014) Food security report 2014 – EU's role to end hunger by 2025. Brussels

Chambers R, Pacey A, Thrupp LA (eds) (1989) Farmer first: farmer innovation and agricultural research. Intermediate Technology Publications/The Bootstrap Press, London/New York, xx + 218 pp

Chappell MJ, Bernhart A et al (2018) Agroecology as a pathway towards sustainable food systems. Available at http://www.db.zs-intern.de/uploads/1540849992-synthesis-report-agroecology.pdf. https://doi.org/10.13140/RG.2.2.12122.59842

Daum T, Birner R (2020) Agricultural mechanization in Africa: myths, realities and an emerging research agenda Glob Food Sec 26:1–10, 100393

Daum T, Villalba R, Anidi O, Mayienga SM, Gupta S, Birner R (2021) Uber for tractors? Opportunities and challenges of digital tools for tractor hire in India and Nigeria. World Dev 144:1–15. https://doi.org/10.1016/j.worlddev.2021.105480

Dawson N (2016) Why the 'green revolution' is making farmers poorer in Rwanda. The Conversation, 17 February. Available at http://theconversation.com/why-the-green-revolution-is-making-farmers-poorerin-rwanda-54768

Dittoh S (1981) Green revolution or evolution? The case of independent African countries. Afr Dev VI(3):8–62

Dittoh S (2003) Improving availability of nutritionally adequate and affordable food supplies at community levels in West Africa. In: Brouwer ID, Traore AS, Treche S (eds) Proceedings of the second international workshop on food-based approaches for a healthy nutrition in West Africa. University of Ouagadougou/IRD/Wagenningen University/FAO. University of Ouagadougou Press, pp 51–62

Dittoh S (2021) Knowledge for service – digital technology positives & negatives in African rural societies. Paper presented in DigHum Panel Series, 23 March 2021. Digital Humanism

Dittoh S, Akuriba MA (2018) Africa's looming food and nutrition insecurity crisis – a call for action. Ghana J Agric Econ Agribus 1(1):148–170

Dittoh S, Omotesho O, Belemvire A, Akuriba M, Haider K (2012) Improving the effectiveness, efficiency and sustainability of fertilizer use in sub-Saharan Africa. Policy Research Paper 3, Global Development Network (GDN): The Global Research Capacity Building Program

FAO (2016) Agricultural mechanization: a key input in for sub-Saharan African smallholders. Integr Crop Manag 23

FAO & AUC (2018) Sustainable agricultural mechanization: a framework for Africa. Addis Ababa, 127 pp. Licence: CC BY-NC-SA 3.0 IGO

Farmer BH (ed) (1980) Green revolution? Technology and change in rice-growing areas of Tamil Nadu and Sri Lanka. Macmillan Press, London and Basingstoke

Feder G, Just R, Zilberman D (1981) Adoption of agricultural innovations in developing countries: a survey. World Bank Staff Working Paper No. 44, Washington, DC

Godfray HCJ, Beddington JR, Crute IR, Haddad L, Lawrence D, Muir JF, Pretty J, Robinson S, Thomas SM, Toulmin C (2010) Food security: the challenge of feeding 9 billion people. Science 327(5967):812–818

Griggs D (2015) From MDGs to SDGs – key challenges and opportunities. Future earth research for global sustainability. Monash Sustainability Institute. https://sustainabledevelopment.un.org/content/documents/3490griggs.pdf. Accessed 06/06/2024

Hazell PBR (2013) Options for African agriculture in an era of high food and energy prices. Agric Econ 44(Supplement 1–5):19–27

Horowitz MM (1990) Donors and deserts: the political ecology of destructive development in the Sahel. Afr Environ VII 1–2–3–4(25–26–27–28):185–209

IPES-Food (2018) Breaking away from industrial food systems: seven case studies of agroecological transition. Available at http://www.ipes-food.org/_img/upload/files/CS2_web.pdf

Johnston BF, Kilby P (1975) Agriculture and structural transformation: economic strategies in late-developing countries. Economic development series. Oxford University Press, 475 pp

Kassie M, Jaleta M, Shiferaw B, Mmbando F, Mekuria M (2013) Adoption of interrelated sustainable agricultural practices in smallholder systems: evidence from rural Tanzania. Technol Forecast Soc Chang 80:525–540. https://doi.org/10.1016/j.techfore.2012.08.007

Kaya HO, Seleti YN (2013) African indigenous knowledge systems and relevance of higher education in South Africa. Int Educ J Comp Perspect 12(1):30–44. ISSN 1443-1475 © 2013 www.iejcomparative.org

Kazembe C (2021) The gap between technology awareness and adoption in Sub-Saharan Africa a literature review for the DeSIRA project. IFPRI Project Note

Kebebe E (2019) Bridging technology adoption gaps in livestock sector in Ethiopia: an innovation system perspective. Technol Soc 57:30–37

Kim J, Mason NM, Snapp S, Wu F (2019) Does sustainable intensification of maize production enhance child nutrition? Evidence from rural Tanzania. Agric Econ 50(6):723–934

Koohafkan P, Altieri MA (2010) Globally important agricultural heritage systems: a legacy for the future. UN-FAO, Rome

Latulippe N, Klenk N (2020) Making room and moving over: knowledge co-production, Indigenous knowledge sovereignty and the politics of global environmental change decision-making. Curr Opin Environ Sustain 42:7–14. https://doi.org/10.1016/j.cosust.2019.10.010

Lele U (1975) The design of rural development: lessons from Africa. Johns Hopkins University Press for the World Bank, Baltimore

Ludwig D (2016) Overlapping ontologies and indigenous knowledge. From integration to ontological self-determination. Stud Hist Philos Sci Part A 59:36–45

Maclean K, Cullen L (2009) Research methodologies for the co-production of knowledge for environmental management in Australia. J R Soc N Z 39(4):205–208. 1175-8899 (Online)

Manuel-Navarrete D, Buzinde C, Swanson T (2021) Fostering horizontal knowledge co-production with Indigenous people by leveraging researchers' transdisciplinary intentions. Ecol Soc 26(2):22. https://doi.org/10.5751/ES-12265-260222

Mapara J (2009) Indigenous knowledge systems in Zimbabwe: juxtaposing postcolonial theory. J Pan Afr Stud 3(1)

Mapemba L, Ward P, Bell A, Kenamu E, Nyirenda Z, Msukwa W (2019) Farmer risk perceptions about conservation agriculture: insights from Malawi. In: 2009 AAAE sixth conference, September 23–26, 2019, Abuja, Nigeria, African Association of Agricultural Economists (AAAE)

Millar D, Dittoh S (2004) Interfacing two knowledge systems: local knowledge and science in Africa. Ghana J Dev Stud. FIDS, UDS 1(2):70–84

Mokuwa A, Nuijten E, Okry F, Teeken B, Maat H et al (2013) Robustness and strategies of adaptation among farmer varieties of African rice (Oryza glaberrima) and Asian rice (Oryza sativa) across West Africa. PLoS One 8(3):e34801. https://doi.org/10.1371/journal.pone.0034801

Moller H, Charleton K, Knight B, Lyver O'B (2009) Traditional ecological knowledge and scientific inference of prey availability: harvests of sooty shearwater (Puffinus griseus) chicks by Rakiura Māori. NZ J Zool 36:259–274

Mudhura M, Critchley W, Di Prima S, Dittoh S, Sessay M (eds) (2016) Community innovations in sustainable land management: lessons from the field in Africa. Earthscan studies in natural resource management. University of KwaZulu Natal, South Africa and Routledge (Taylor and Francis Group), London

Muya S (2007) The efficacy of traditional knowledge in African education. Paper presented at the conference on education and Indigenous knowledge systems in Africa. College of Business Education, Dar es Salaam, Tanzania, 2–3 August

Navarro S, Donahaye E, Fishman S (1994) The future of hermetic storage of dry grains in tropical and subtropical climates. In: Highley E, Wright EJ, Banks HJ, Champ BR (eds) Proceedings of the 6th international working conference on stored-product protection, Canberra, Australia, 17–23 April 1994. CAB International, Wallingford, Oxon, pp 130–138

Offei SK, Almekinders C, Crane TA, Hughes SG, Mokuwa A, Nuijten E, Okty F, Struik PC, Teenken B and Richards P (2010) Making better seeds for Africa food security - a new approach to scientist-farmer partnership. *Aspects of Applied Biology* 96

Petersen B, Snapp S (2015) What is sustainable intensification? Views from experts. Land Use Policy 46:1–10

Pingali PL (2007) Agricultural mechanization: adoption patterns and economic impact. Handbook Agric Econ 3:2779–2805. https://doi.org/10.1016/S1574-0072(06)03054-4

Pingali PL (2012) Green revolution: impacts, limits, and the path ahead. Proc Natl Acad Sci USA 109:12302–12308. https://doi.org/10.1073/pnas.0912953109

Pretty J, Toulmin C, Williams SB (2011) Sustainable intensification in African agriculture. Int J Agric Sustain 9(1):5–24

Sheahan M, Barrett CB (2017) Ten striking facts about agricultural input use in Sub-Saharan Africa. Food Policy 67:12–25. https://doi.org/10.1016/j.foodpol.2016.09.010

Teeken B, Nuijten E, Temudo MP, Okry F, Mokuwa A, Struik PC, Richards P (2012) Maintaining or abandoning African rice: lessons for understanding processes of seed innovation. Hum Ecol 40:879–892. https://doi.org/10.1007/s10745-012-9528-x

Tilman D, Cassman KG, Matson PA, Naylor R, Polasky S (2002) Agricultural sustainability and intensive production practices. Nature 418:671–677. https://doi.org/10.1038/nature01014

Toledo VM (1990) The ecological rationality of peasant production. In: Altieri M, Hecht S (eds) Agroecology and small farmer development. CRC Press, Boca Raton, pp 51–58

UNDP and World Bank (2015) Transitioning from MDGs to SDGs. UNDP/World Bank, New York/Washington DC, 176 pp. undp.org. Accessed 06/06/2024

Veitayaki J (1997) Traditional marine resource management practices used in the Pacific Islands: an agenda for change. Ocean Coast Manag 37(1):123–136

Williams PA, Sikutshwa L, Shackleton S (2020) Acknowledging Indigenous and local knowledge to facilitate collaboration in landscape approaches—lessons from a systematic review. Land 9(9):39. https://doi.org/10.3390/land9090331

Wise TA (2020) Failing Africa's farmers: an impact assessment of the alliance for a green revolution in Africa. Available at https://sites.tufts.edu/gdae/files/2020/10/20-01

Roundabout Journey to Resilient and Sustainable Food Systems in West Africa

Saa Dittoh, Mojisola Olanike Kehinde, and Anna Bon

Abstract

The Chapter reviews the "agricultural development paths" in West Africa from the 1950s, using Nigeria and Ghana as case studies. It notes that the smallholder farmer production practices in the 1950s and 1960s were generally resilient and sustainable and the countries were food and nutrition secure. Africa's food insecurity dilemma has been mainly due to the shift of focus to export crops, combined with inappropriate "agricultural modernization" methodologies, and globalization, as well as the sidelining of indigenous knowledge of food systems. The result has been non-resilient and unsustainable food systems, increasing food and nutrition insecurity and negative environmental consequences. Based on historical and field research information, we argue that, to ensure production of adequate, healthy, safe and nutritious foods for all people, especially in West Africa, a shift must be made back to indigenous and local African knowledge of food production, preservation, processing and consumption, but enriched with new and appropriate environmentally friendly and adaptive knowledge systems, such as agroecology, which has its roots in indigenous production systems. The appropriate blending of traditional and new knowledge that is based on "responsible science" (science that enhances the elements of the natural environment and people) is what the plug-in principle is all about.

Keywords

Smallholder farmers · West Africa · Green revolution · Plug-In · Nature positive food systems · Agroecology

2.1 Introduction

The agricultural underdevelopment and food insecurity story in Africa and particularly West Africa is very pathetic. Most African countries were food secure in the precolonial times, despite the fact that the slave trade of the 1500s had negative impact on food production in Africa (Oculi 1981). It has been stated that slave trade records of the 1800s suggest that Africans were well-nourished, strong and of regular stature (Diffie

S. Dittoh (✉)
University for Development Studies, Tamale, Ghana

M. O. Kehinde
Landmark University,
Omu-Arun, Kwara State, Nigeria

A. Bon
Vrije Universiteit Amsterdam,
Amsterdam, Noord-Holland, The Netherlands

© The Author(s) 2025
S. Dittoh et al. (eds.), *Integrating Indigenous and Scientific Knowledge for Sustainable Food Systems in Africa*, Sustainable Development Goals Series,
https://doi.org/10.1007/978-3-031-85512-2_2

et al. 1977; Hawthorne 2003), and up to the 1950s and 1960s sub-Saharan Africa (SSA) was food secure (Bjornlund et al. 2020, 2022). According to Olayemi and Dittoh (1995), "the neglected food economy (of Nigeria) performed sufficiently well to produce enough food for the populace" in the early 1960s. The method of production was traditional mixed cropping and mixed farming of staple food crops and local breeds of livestock and poultry as well as use of farmyard manure and traditional soil management and conservation methods. It was the same in almost all West African countries.

The genesis of Africa's food insecurity can be traced to colonialism and the introduction of capitalist model of production (Paitnak 1972; Shepherd 1981). After the abolition of the slave trade in the early nineteenth century, commerce shifted from slave trade to agriculture and the production of export crops. Colonial governments encouraged and supported the production of export crops: oil palm, groundnuts, cocoa, cotton, tea, coffee, tobacco, rubber, sisal and others for industries in Europe and North America (Oculi 1981; Roessler et al. 2022), to the neglect of the staple food crops production in the colonies. By the 1950s, agricultural commodities accounted for 65% of total exports across 38 African states (Hance et al. 1961). There was overt and covert suppression of the African food production systems (of mixed cropping and mixed farming) and focus was on export crops for the then booming agricultural export trade (Hopkins 1973; Tosh 1980; Austin 2014; Frankema et al. 2018). Even though plantation-oriented agriculture did not take root in West Africa, as was in Latin America, smallholder farmers in West Africa concentrated on the production of cash crops (Hogendorn 1969; Curtin 1990).

After political independence in late 1950s and early 1960s, many African governments continued the agricultural export model of "development" and continued to emphasise production for export with lip service paid to staple food production. The emphasis on these internationally traded crops (and minerals such as gold, diamonds and others) became even greater with the rise of transnational corporations (TNCs) and globalization (Khor 2000). The growth of globalization went hand in hand with the growth of international agribusinesses and the promotion of high yielding crop varieties (HYVs) (especially maize and rice), fertilizers and agrochemicals. The catch phrase in many African countries, was "agricultural modernization" and it was mainly about adoption of HYVs and their complementary inputs, fertilizers and agrochemicals, and the use of tractors. The thrust of most agricultural development policies, plans, strategies, projects and programmes of most African governments and donor agencies, including the World Bank, the African Development Bank and others up till now is "agricultural modernization" which they believe is the "new green revolution" (World Bank 2007; MOFA 2010; AfDB 2016; AGRA 2017). Even though there is now some realization of the misplaced priorities with respect to green revolution (GR) technologies, policies and strategies are yet to change significantly in most African countries. The irony is that the more "agricultural modernization" is pushed, the greater the food and nutrition insecurity in African countries. In Ghana, for example, per capita food production declined at the rates of 3.1%, 3.5% and 1.3% respectively for 1970–75, 1976–80 and 1981–85 while for Nigeria the per capita food production stagnated at 0.3 for the whole three periods (Balogun 1992), and Nigeria that was a leading producer and exporter of palm oil in the 1950s and 1960s became a net importer of vegetable oil (Balogun 1992).

It is important to note that there have been serious food insecurity situations (famines) since 1960 in parts of Africa especially in the 1970s because of natural disasters such as droughts and/or civil conflicts, but as stated by Bjornlund et al. (2022) the situations were "not due to unavailability or insufficient regional production of food. Rather, they arose from governments' failure to integrate and keep track of rural food production, storage and distribution systems". That means governments did nothing to support what the farmers and communities could have done, using their indigenous knowledge and methods, to stem the food insecurity. It is well documented that rural people in Africa and elsewhere have numerous resilience and coping strategies against food insecurity which they use to lessen the effects of food insecurity. They are able to use the strategies effectively in times of disasters and conflicts, especially if there is some government or external assistance (Amendah et al. 2014; Amfo et al. 2021; Danso-Abbeam et al. 2023; Biadgilign 2023; Gebrehiwot et al. 2024).

It is now argued that the industrial food system (the green revolution model) has been the major driver of the current broken world food system. Ariga et al. (2019) undertook a rigorous assessment of the ability of technologies embodied in improved seeds and chemical fertilizers to stimulate a green revolution in Africa and found that they cannot for various reasons. Also, according to von Braun et al. (2021), "the world's food system is in disarray, one in ten people is undernourished, one in four is overweight and more than one-third of the world's population cannot afford a healthy diet". Others have pointed out that the GR approach resulted in groundwater depletion, soil degradation, loss of biodiversity, and water pollution (Shiva 1991; Murgai et al. 2001; Pretty et al. 2011), and that the focus on a few dominant cereals adversely affected the production of nutritionally important crops (Pingali 2012). The latter situation is very visible in Africa where hybrid maize, rice and soybean have replaced large areas of land where millet, sorghum, cowpea, groundnuts, and various roots and tubers, which are all nutritionally superior, would have been cultivated. Farmers have also continued to complain of the high cost of the modern technologies leading the marginalization and impoverishment of numerous smallholder farmers (Hazell 2009; Pingali 2012). The goal of ending hunger, achieving food security and improved nutrition, and promoting sustainable agriculture by 2030 seems to be an impossible task unless there is a drastic change in agricultural development strategy, especially in Africa.

How did we get here with all the science and technology research and innovations and the "massive investments" by African governments and development partners (DPs) in Africa's agricultural sector over the years? This Chapter traces the various attempts of food and agricultural development in West Africa from the 1960s and 1970s to the present, with emphasis on the situations in Nigeria and Ghana. It analyses the "agricultural development paths" of Nigeria and Ghana to show their unsustainability and lack of resilience and the fact that the "agricultural modernization" models that they have pursued over time and even still pursuing will continue to perpetuate food and nutrition insecurity, poverty of farming families and communities, and environmental and ecosystems services deterioration. The main aim is to suggest how research and/or policy can leverage the knowledge and potential of smallholder farmers to ensure resilient and sustainable food systems in the sub-continent. The argument is that if there had been "plug-in" by international and national agricultural institutions into what smallholder farmers were doing, resilient and sustainable food systems would have been achieved a long time ago. There would not have been the need to be talking about food systems transition in 2024. There has been an unnecessary roundabout journey for decades. A "plug-in" to local farmers' practices with appropriately designed "modern technology" was all that was required.

2.2 Neglect of Small Farmer Model and the Replacement Mentality

According to M. S. Swaminathan, a former Director-General of the International Rice Research Institute (IRRI), who is often referred to as the Father of the Green Revolution in India, "the world's most experienced farmers are found in the Third World" (Swaminathan 1983). Also, a former Director-General of the International Institute of Tropical Agriculture (IITA) said "all evidence points to the fact that (African) smallholders are outstanding managers of their own resources and can be counted on to accept changes in farm operations that promote greater profitability" (Hartman 1984). Pattanayak (1983) also reported that attempts in 1975 by the

International Crops Research Institute for the Semi-Arid Tropics (ICRISAT) to introduce high yielding varieties of sorghum and millet into West Africa failed and that recorded yields were far lower than that of local varieties. Again, IITA's assessment of its improved maize technology in South-East Nigeria in 1975, showed that farmers would have been worse off if they adopted the capital-intensive technology than their existing relatively low-cost system of production (IITA 1976). The critical question that arises is why are these outstanding and experienced managers of farms (that is, smallholder farmers) unable to produce enough food to feed the African population especially as their forefathers were able to do so in the eighteenth and nineteenth centuries? Also, why are these facts about small farmer production not leveraged upon by agricultural research and/or government and donor policies? Why will a World Bank Vice President say in 1984 that "… we and everybody else are still unclear about what can be done in agriculture in Africa" (Stern 1984), given all the experiences gained at the various international, regional and national research centres about the ingenuity of the African smallholder farmer?

Several researchers and analysts of the African agrarian problems have called for a green revolution of the Asian type (Aribisala 1983; Jama and Pizarro 2008; Toenniessen et al. 2008; Breisinger et al. 2011) resulting in the establishment of the Alliance for Green Revolution in Africa (AGRA). Others have argued against such kind of a green revolution for Africa and suggested an *Agrarian Evolution* (Dittoh 1981), a *Grey to Green Revolution* (Dar et al. 2004) and an *Evergreen Revolution* (Swaminathan 2006). Several others have put Africa's agricultural problem at the doorsteps of lack of viable extension services (Okigbo 1985), and unchecked population growth (World Bank 1981; Eicher 1985; Brown and Wolfe 1988; Stifel 1986). What is noticeable in all these arguments is that the views do not explicitly consider the smallholder farmers' opinion. They all sideline the knowledge and views of smallholder farmers, maybe with the erroneous view that they are "conservative, ignorant and superstitious" (Aribisala 1983). There

are several people who argue that smallholder farmers are not "commercial", that is, they are largely subsistence (Govereh et al. 1999; IFAD-IFPRI 2011; Gebreslassie et al. 2015; AGRA 2017), and yet they produce 90% of the food consumed by both rural and urban people in Africa. According to several sources, smallholder farmers (cultivating 2 ha or less) produce 70–80% of the world's food (ETC 2009; FAO 2014; Ricciardia et al. 2018). Others argue that smallholder farmers should transition to medium size (Jayne and Ameyaw 2016) and then maybe to large size. The opposite is, however, happening in several places across the globe; there is a large body of research evidence of the inverse relationship of size and productivity (Beets 1982; Dyer 1991). Very little deep analyses exist on the views of farmers and stakeholders at the grassroots on alternative paths to resilient and sustainable food systems (Dorward 1999; Rosset 1999). All the wrong beliefs stated by researchers arise from the wrong approach to research and to interventions which refuse to start from the known to the unknown. Too many researchers seem to have conclusions before starting their research. They undertake research to get what they have already decided is the result. That obviously is not research.

There has been systematic sidelining and neglect of smallholder farmers' invaluable knowledge and practices, even though there is considerable evidence that resilient and sustainable food production systems reside in their practices. As stated by Uma Lele, a renowned agricultural economist and a former Senior Advisor to the World Bank in 1981, "African countries are not giving priority to the development of peasant (smallholder) agriculture. There is not even much understanding of what is required to develop it" (Lele 1981). It is obvious that after over four decades, there is still not much understanding of what is required to develop African smallholder agriculture. The industrial agricultural production system with its surrogate wrong agribusiness models continues to be used by African Governments and Development Partners (DPs) with continued believe in "replacement", as opposed to the integration, of systems. George Benneh, a renowned African scholar, also said there is the need for African researchers "to fish from the indigenous creative pool of ideas to produce low-cost appropriate technologies which are suitable for the environmental conditions of African countries" (Benneh 2011).

2.3 Food and Agricultural Development Paths in West Africa from the 1960s

As explained earlier, West Africa, and indeed all of Africa, was relatively food sufficient, and food secure up till the early 1960s. Transient and seasonal food shortages did occur in places but not on a scale that could be regarded as states of food insecurity. West African countries began to experience food shortages in the 1970s mainly through the long period of neglect of the local food sector, globalization of food production and distribution (Vidal 2011; Kaufman 2010; Hari 2010) and rapid population increases. By the 1980s many of the countries were "hungry, heavily indebted, marginalised nations depending on the whims and caprices of industrialized nations, the World Bank and the IMF" (Dittoh 1991).

Attempts at tackling the food and agriculture problem did start by governments, development agencies and non-government organizations in the 1960s. The interventions were, however, more about the introduction of fertilizers and improved seeds to farmers. It was a precursor to the "agricultural modernization" model. Governments and the various development agencies used various entry points such as subsidies, credit programmes, control of output markets among others to stimulate fertilizer demand (Morris et al. 2007). They also ensured constant and adequate supply of the inputs. That stimulated productivity and even profitability to the farmer. Those interventions led to upscaling and out-scaling of "green revolution" practices.

In Nigeria, the World Bank-assisted agricultural development projects (ADPs) were started in the northern parts of the country in the early 1970s and were "the core of Nigeria's Green Revolution programme" (D'Silva and Raza 1983). They were established in 1975 as enclave projects in Gusau (Sokoto State), Funtua (Kaduna State) and Gombe (Bauchi State).[1] More were established in seven other states between 1977 and 1981 while the earlier ones were upgraded to state-wide ADPs. By 1989, all the then 19 States of Nigeria had state-wide ADPs (Weaving 1993). The World Bank loaned Nigeria US$1.2 billion between 1975 and 1983 towards the ADPs (*African Business*, July 1985 and January 1986). The ADP model was a top-down supply-driven one which was initiated, planned and executed by the World Bank, the Federal Government and State Government bureaucrats. Local government personnel were bypassed, and farmers (progressive farmers) were only "receivers of international technology transfer (ITT)" (Dittoh and Akatugba 1988). According to Weaving (1993) and the World Bank (1995) the projects "relied heavily on expatriate consultants". The projects targeted progressive farmers with a belief that the transferred technology would "trickle down" to other (poorer) farmers.

Dittoh and Akatugba (1988) compared yields before and after the three early enclave projects (Funtua, Gusau and Gombe). The information is as given in Tables 2.1, 2.2, and 2.3. The yields information given in the tables clearly indicates that yield differences were not significant despite massive expenditures on the three enclave projects for about 5 years. If the environmental deterioration as a result of heavy machines used on the fragile soils, and the heavy doses of highly subsidized fertilizers are taken into consideration, the interventions could be described as major failures. An evaluation of the ADPs by the World Bank in 1993 indicated their poor performance (Weaving 1993).

The most obvious indication of the failure of the ADPs is their unsustainability and lack of resilience. Hardly any benefits can be traced to the originally designed projects. A multidisci-

---

[1] Over the years many States have been created out of the then existing States. Funtua is now located in Katsina State, Gusau is in Zamfara State and Gombe is in Gombe State.

Table 2.1 Yields of some crop mixtures in Zaria (near Funtua) in 1966/67 (before the establishment of the ADPs)

Crop	Crop mixture		
	Millet/sorghum (Yield: kg/ha)	Millet/sorghum/cowpea (Yield: kg/ha)	Average yield (kg/ha)
Funtua	370	400	385
Sorghum	768	714	741
Cowpea	–	167	167

Source: Norman et al. (1982), p. 167

Table 2.2 Yields of some crops mixtures in Sokoto (near Gusau) in 1966/67 (before the establishment of the ADPs)

Crop	Crop mixture		
	Millet/sorghum (Yield: kg/ha)	Millet/sorghum/cowpea (Yield: kg/ha)	Average yield (kg/ha)
Millet	892	772	832
Sorghum	186	124	155
Cowpea	–	63	63

Source: Norman et al. (1982), p. 167

Table 2.3 Average yields of some crops in Funtua ADP (near Zaria) and Gusau ADP (near Sokoto) in 1979/80 (Projects year 5)

Crop	Funtua (near Zaria) ([a]Average yield: kg/ha)	Gusau (near Sokoto) ([a]Average yield: kg/ha)
Millet	427	587
Sorghum	738	489
Cowpea	104	106

Source: Okorie and Idachaba (1983), p. 66
[a]Average yield refers to the yield of a crop in all mixtures including sole crops
Sole cropping was encouraged

plinary team undertook an evaluation of the ADPs in 2008 and arrived at the conclusion that, "the Agricultural Development Programs from Sokoto to Calabar and Maiduguri to Lagos, are becoming shadows of their past; this is indeed a sad story for Nigerian Agriculture" (Auta and Dafwang 2010).

Another agricultural development strategy that the Federal Government of Nigeria implemented almost side by side with the ADPs was through 11 River Basin Development Authorities (RBDAs) in 1976 which were increased in 1984 to 18 River Basin and Rural Development Authorities (RBRDAs). They were established throughout the country for the purpose of exploiting water resources for multipurpose uses (Balogun and Ukeje 1986) and to accelerate Nigeria's agricultural and rural development (Gana et al. 2019). They were described as the linchpin of Nigeria's agricultural develop-

ment (*African Business*, November 1981). They were funded by the Federal Government and functioned as separate authorities without the State and local governments playing any role in their activities. Just as the ADPs, they were top-down supply-driven establishments, characterized by large/medium scale capital-intensive dams and irrigation infrastructures, land clearing and input distribution to farmers (Balogun and Ukeje 1986; Salau 1986). Gana et al. (2019) evaluated the performance of the RBRDAs and concluded that they "were poorly organized ... and failed to boost food and industrial crop production as expected" and Adams (1985) described the economic performance and social impacts of the schemes as disappointing. Balogun and Ukeje (1986) were more specific in the areas of failure by stating that "the super-imposition of the sprinkler irrigation system on a labour-intensive agriculture resulted in the over-capitalisation and under utilisation of both human and capital resources", while Akindele and Adebo (2004) blamed their failure on political interference.

We believe the real reason for the failure of both the ADPs and RBRDAs was the fact that they were based on poorly conceived green revolution (GR) and international technology transfer (ITT) models and had very limited chance of success. It has been shown and argued, since the 1970s that agricultural growth based on GR technologies and ITT are inherently unsustainable and are bound to fail (Farmer 1980; Dittoh 1981; Ninan and Chandrashekar 1992; McIntyre et al. 2009; FAO 2016; UNEP 2016; von Braun et al. 2021). GR and ITT models are designed to "replace" existing smallholder farmers' technologies. They do not attempt to "integrate" whatever technologies are being transferred with local knowledge and practices. According to John and Babu (2021), "many interventions, beneficial for the shorter term, such as the green revolution, without the consideration of ecological principles, can be detrimental and irreversible in the long run".

Lessons learnt from the failures of the ADPs and the RBRDAs might have been the reason a type of "plug-in" participatory approach was used in the establishment of Nigeria's World Bank assisted National Fadama Development Project (Mohan 2002; Hima et al. 2016). According to Hima et al. (2016), an "inclusive and participatory model of local economic decision making was used", and there was "strong beneficiary involvement in project design and implementation" which reinforced "a sense of ownership, including responsibility for the maintenance of infrastructural facilities" (Mohan 2002). It has been argued that the key to success in a rural development program is the extent to which local people are involved in its planning and implementation (World Bank 1975; Chirenje et al. 2013; Kinyata and Abiodun 2020). There are indications that the Fadama model is largely sustainable. The relative success of the Fadama projects must have been due to infusions of local knowledge and practices by the farmers. The ADP and RBRDA initiatives should have started in like manner by incorporating indigenous farm-

ing knowledge and strongly involving the people. Nigeria's agricultural development journey has been clearly a roundabout journey. It is hoped that the lessons of the Fadama project will inform present and future attempts at the development of rainfed agriculture in Nigeria.

In Ghana, World Bank assisted agricultural and rural development projects started with the Upper Regional Agricultural Development Programme (URADEP) in 1976, followed by the Volta Regional Agricultural Development Programme (VORADEP) and others in the 1970s and 1980s (Chambers 1980; Agunga 1983). Several other DPs such as IFAD, GTZ/GIZ, CIDA, USAID, AFD, the AfDB and others as well as NGOs also got involved and several donor-funded agricultural projects have been implemented over several decades. The target areas of most of the interventions have been the transitional and northern agroecological zones of Ghana. A few examples of the interventions include Northern Rural Growth Project (NRGP), Savannah Zone Agricultural Productivity Improvement Project (SAPIP), Ghana Agricultural Sector Investment Programme (GASIP), Rural Technology Improvement Project (RTIP), and others. In all the numerous projects and programmes it is difficult to point to any sustainable achievements, especially given the level of resources expended (Agunga 1983; Aryeetey 1990; Kijima et al. 2011). The same top-down industrial agricultural GR and ITT models are used in almost all the projects. However, as pointed out by Robert McNamara, one time President of The World Bank, "no program will help small farmers if it is designed by those who have no knowledge of their problems and operated by those who have no interest in their future". (World Bank 1975). That seems to be the situation in several donor assisted projects. There has hardly been any incorporation of smallholder farmers' knowledge in the interventions. Their local knowledge and expertise continue to be ignored, and the interventions continue to achieve very little.

Thus, every strategy of agricultural development in Ghana since the 1970s has been along the models of GR and ITT, as in Nigeria's ADP and

RBRDA cases. It has been about replacing "traditional agriculture" with "modern agriculture". Even with the realization that realistic modernization can be done through smallholder farmers (World Bank 2009), GR and ITT models are still the focus. The efforts have not been to improve smallholder farmer systems but to replace them. The smallholder farmer will operate efficiently only in systems he/she understands. Smallholder farmers are "systems thinkers and actors". Models that ignore systems concepts are unlikely to resonate with African smallholder farmers. Conventional GR and ITT models have continued to sideline indigenous knowledge and indigenous agricultural practices, which are largely systems oriented. It is now obvious that to achieve resilient and sustainable food systems there is a need to get back and integrate "modern science" with "traditional science (ethnoscience)" in a systems way. GR and ITT models can no longer ignore indigenous knowledge. Good "plug-in" integration of modern and indigenous science is the path to resilient and sustainable food systems not only in Africa but around the globe.

The main difference between the "agricultural development paths" of Ghana and Nigeria since the 1970s is that the Federal Government of Nigeria had been funding some agricultural development projects such as RBRDA projects and others. In the case of Ghana. almost all agricultural interventions have been donor-funded and donor-driven. According to the World Bank (2017), DPs fund more than 70% of Ghana's agriculture investment expenditures and most of the "agricultural public expenditure is for input subsidies". That means apart from input subsidies, which most DPs have refused to fund, Ghana government direct investment in the food sub-sector has been almost nil. According to Kolavalli et al. (2008), Ghana's government expenditure on Research and Development (R&D) and for the agricultural sector for several years has been for only recurrent activities, mainly salaries and wages. It is similar in many other African countries (Benin and Yu 2013).

2.4 Towards Resilient and Sustainable Food Systems

2.4.1 Nature Positive Food Systems

The evidence in all parts of the world is that the prevailing global industrial food system is not providing adequate, healthy, safe and nutritious foods for the people worldwide (FAO 2016; WFP 2017). The system is also environmentally and ecosystems services destructive and very vulnerable to climate change effects. There is a need to move towards more sustainable and resilient agricultural and food systems which depend more on available local resources (Welch and Graham 2000; Caritas Europa 2014; FAO 2016). A resilient and sustainable food system should be able to deliver food and nutrition security for all in the short, medium and long term.

A food system is often defined as "the entire range of activities involved in the production, processing, marketing, consumption and disposal of goods that originate from agriculture, forestry, and fisheries, including the inputs needed and the outputs generated at each of the steps (HLPE 2017). A sustainable food system is, therefore, one that ensures environmental, social and economic sustainability at every level along the food chain. Certain key elements are associated with sustainable food systems. They include resilience (ability to withstand shocks); climate smart production; adequate planetary health; ecosystems and biodiversity enhancement; nutrition and gender sensitive agriculture; production of adequate, desirable and safe food; viable, socially acceptable and environmentally friendly value chains; provision of healthy diets; healthy and sustainable consumption patterns and equitable livelihoods (living no one behind). These elements are not mutually exclusive. They are indeed mutually reinforcing and interwoven.

It looks impossible for any known food system to have all the above stated elements, but there are technologies and systems that work towards having most of them. They include the commonly known "sustainable production technologies" such as climate smart agriculture (CSA), sustainable land management (SLM), regenerative agriculture, permaculture, farmer managed natural regeneration (FMNR), agroforestry systems and others, which address one or a few of the elements. Most of them are nature-based, nature-derived and/or nature-inspired practices that offer solutions to specific agricultural problems. They, thus, resonate well with farmers' indigenous knowledge and practices which are largely nature-based. Almost all the above-mentioned technologies are, however, "stand alone" technologies. They are largely discipline and/or subject-based: soil-based, agronomy-based, forestry-based, engineering-based and so on, and do not function as "systems". They are therefore not easily adopted by smallholder farmers, since such farmers are generally "systems thinkers" and "systems actors". It is difficult to incorporate discipline-based (and commodity-based) technologies into their integrated systems. That is why agroecology tends to be a better system since it is "systems-oriented". Also, agroecology research is necessarily "action research" with considerable interdisciplinarity and/or transdisciplinarity (Roling and Jiggins 2009), and the goal of the research is co-creation of technologies and co-innovation with farmers.

Agroecology may be defined as a systems science (the ecology of food systems), a practice (practical combination of knowledge systems) and a movement (mobilization of farmers and other stakeholders for their collective good); which incorporates activities of the whole agroecosystem (including its socio-cultural, economic and political dimensions) not just parts of it. The numerous environmental, social and economic sustainability problems of industrial agriculture and conventional food systems are solved with agroecological practices. Land degradation, green gas emissions, complex land tenure, access to productive resources by women and youth, marketing and income distribution, and several other problems are much easier addressed in agroecological production systems. If "responsible science", science that enhances the elements of the natural environment and people, is used to improve a local or indigenous agroecological system, the result will be resilient and sustainable food systems. Local agroecological practices integrated with "responsible modern science" ensures that co-created technologies are produced and are "systems-oriented".

References

Adams WM (1985) River basin planning in Nigeria. Appl Geogr 5(4):297–308

AfDB (African Development Bank) (2016) Feed Africa: strategy for agricultural transformation in Africa 2016–2025. African Development Bank, Abidjan

AGRA (Alliance for Green Revolution in Africa) (2016) Africa agriculture status report: progress towards agriculture transformation in sub-Saharan Africa. Issue No 4

AGRA (2017) Africa agriculture status report: the business of smallholder agriculture in sub-Saharan Africa. Alliance for a Green Revolution in Africa (AGRA). Issue No. 5, Nairobi

Agunga RA (1983) The World Bank, new directions, and the upper region agricultural development project in Ghana: A case study. MA thesis. University of Iowa, Iowa City

Akindele ST, Adebo A (2004) The political economy of River Basin and rural development authority in Nigeria: A retrospective case study of Owena-River Basin and Rural Development Authority (ORBRDA). J Hum Ecol 16(1):55–62

Amendah DD, Buigut S, Mohamed S (2014) Coping strategies among urban poor: evidence from Nairobi, Kenya. PLoS One 9(1)

Amfo B, Aidoo R, Osei Mensah J (2021) Food coping strategies among migrant labourers on cocoa farms in southern Ghana. Food Secur 13(4):875–894

Aribisala TSB (1983) Nigeria's green revolution: achievements, problems and prospects. Distinguished Lecture No.1. Nigerian Institute of Social and Economic Research (NISER), Ibadan

Ariga J, Mabaya E, Waithaka M, Wanzala-Mlobela M (2019) Can improved agricultural technologies spur a green revolution in Africa? A multicountry analysis of seed and fertilizer delivery systems. Agric Econ 50(S1):63–74

Aryeetey E (1990) Decentralization for rural development: exogenous factors and semi-autonomous programme units in Ghana. Community Dev J 25(3):206–214

Austin G (2014) Explaining and evaluating the cash crop revolution in the "peasant" colonies of tropical Africa, ca. 1890-ca. 1930: beyond "vent for surplus". In: Akyeampong E, Bates RH, Nunn N, Robinson J (eds) Africa's development in historical perspective. Cambridge University Press, New York, pp 295–320

Auta SJ, Dafwang II (2010) The Agricultural Development Projects (ADPs) in Nigeria: status and policy implications. Res J Agric Biol Sci 6(2):138–143

Balogun EA (1992) Agriculture and government policies in West Africa: evidence from Cote D'Ivoire, Ghana and Nigeria. West Afr Econ J 6:68–84

Balogun ED, Ukeje EU (1986) The impact of River Basin development authorities on Nigerian agriculture: A case study of Niger River Basin development authority. CBN Econ Financ Rev 24(2):64–76

Beets WC (1982) Multiple cropping and tropical farming systems. Westview Press, Boulder

Benin S, Yu B (2013) Complying with the Maputo declaration target: trends in public agricultural expenditures and implications for pursuit of optimal allocation of public agricultural spending. ReSAKSS annual trends and outlook report 2012. International Food Policy Research Institute, Washington, DC

Benneh G (2011) Technology should seek tradition – studies on traditional land tenure and smallholder farming systems in Ghana. Ghana Universities Press, Accra

Biadgilign S (2023) Coping strategies to mitigate food insecurity at household level: evidence from urban setting in Addis Ababa, Ethiopia. Inquiry 60:1–11

Bjornlund V, Bjornlund H, van Rooyen AF (2020) Why agricultural production in sub-Saharan Africa remains low compared to the rest of the world – a historic perspective. Int J Water Resour Dev 36(1):S20–S53

Bjornlund V, Bjornlund H, van Rooyen A (2022) Why food insecurity persists in sub-Saharan Africa: A review of existing evidence. Food Secur 14:845–864

Breisinger C, Diao X, Thurlow J, Al Hassan RM (2011) Potential impacts of a green revolution in Africa – the case of Ghana. J Int Dev 23(1):82–102

Brown LR, Wolfe FC (1988) Reversing Africa's decline. Worldwatch Paper 66. Worldwatch Institute, Washington, DC

Caritas Europa (2014) Food security report 2014 – EU's role to end hunger by 2025, Brussels

Chambers MI (1980) The politics of agricultural and rural development in the Upper Region of Ghana: implications of technocratic ideology and non-participatory development. PhD dissertation. Cornell University, Ithaca

Chirenje LI, Giliba RA and Musamba (2013) Local communities' participation in decision-making processes through planning and budgeting in African countries. Chinese Journal of Population Resources and Environment. 11(1):10–16. http://dx.doi.org/10.1080/10042857.2013.777198

Curtin PD (1990) The rise and fall of the plantation complex: essays in Atlantic history. Cambridge University Press

D'Silva BC, Raza MR (1983) Equity considerations in planning and implementing rural development projects in Nigeria: an evaluation of the Funtua project. In: Greenshields BL, Bellamy MA (eds) Rural devel-

opment: growth and inequity. IAAE Occasional Paper No. 3

Danso-Abbeam G, Asale MA, Ogundeji AA (2023) Determinants of household food insecurity and coping strategies in Northern Ghana. GeoJournal 88:2307–2324

Dar WD, McGaw EM, Reddy DVR (2004) Chapter 5: A grey to green revolution in the semi-arid tropics of Asia and Africa. In: Rao SC, Ryan J (eds) Challenges and strategies in dryland agriculture. Crop Science Society of America (CSSA) Special Publication 32, Madison, pp 35–46

Diffie BW, Shafer BC, Winius GD (1977) Foundations of the Portuguese empire, 1415–1580. University of Minnesota Press

Dittoh S (1981) Green Revolution or evolution? The case of independent African countries. Afr Dev VI(3):8–62

Dittoh S (1991) The debt crisis and agricultural development in Nigeria in the 1990s and after. In: Olowo T, Akinwumi JA (eds) Proceedings of the national conference of the Ibadan socio-economic group on development strategies in 21st century Nigeria. African Economic Research Consortium Inc

Dittoh S, Akatugba D (1988) The politics and economics of food policy planning and implementation in Nigeria. In: Sanda AO (ed) Corporate strategy for agricultural and rural development in Nigeria. Obafemi Awolowo University, Ile-Ife

Dorward A (1999) Farm size and productivity in Malawian smallholder agriculture. J Dev Stud 35:141–161

Dyer G (1991) Farm size – farm productivity re-examined: evidence from rural Egypt. J Peasant Stud 19(1):59–92

Eicher CK (1985) Famine prevention in Africa: the long view. In: Food for the future: proceedings of the bicentennial forum. Philadelphia Society for the Promoting Agriculture, Philadelphia

ETC-Group (2009) Who will feed us? Questions for the food and climate crises

FAO (2014) The state of food and agriculture 2014: innovation in family farming Food and Agriculture Organization of the United Nations

FAO (2016) The state of food and agriculture: climate change. Agriculture and Food Security, Rome

Farmer BH (1980) (Ed.) Green Revolution? Technology and Change in Rice-growing Areas of Tamil Nadu and Sri Lanka. Macmillan Press, London and Basingstoke

Frankema E, Williamson J, Woltjer P (2018) An economic rationale for the west African scramble? The commercial transition and the commodity price boom of 1835–1885. J Econ Hist 78(1):231–267

Gana BA, Abdulkadir IF, Musa H, Tijjani Garba T (2019) A conceptual framework for organization of River Basin development and management in Nigeria. Eur J Eng Res Sci (EJERS) 4(6):34–40

Gebrehiwot M, Demisse B, Meaza H, Gezahegn TW, Abbay AG, Beyene S (2024) Understanding food security determinants and coping strategies among smallholder farming households in Northern Ethiopia. Cogent Soc Sci 10(1). https://doi.org/10.1080/23311886.2024.2354970

Gebreslassie H, Kebede M, Kiros-Meles A (2015) Crop commercialization and smallholder farmers' livelihood in Tigray region, Ethiopia. J Dev Agric Econ 7(9):314–322

Govereh J, Jayne TS, Nyoro J (1999) Smallholder commercialization, interlinked markets and food crop productivity: cross-country evidence in Eastern and Southern Africa. Department of Agricultural Economics and the Department of Economics, Michigan State University, East Lansing

Hance WA, Kotschar V, Peterec RJ (1961) Source areas of export production in Tropical Africa. Geogr Rev 51(4):487–499

Hari J (2010) How Goldman gambles on starvation. The Independent, 2 July

Hartman EH (1984) Prospects for Nigerian agriculture. Distinguished Lecture No.2. Nigerian Institute of Social and Economic Research (NISER), Ibadan

Hawthorne W (2003) Planting rice and harvesting slaves: transformations along the Guinea-Bissau Coast, 1400–1900. Heinemann

Hazell PBR (2009) The Asian green revolution. International Food Policy Research Institute, discussion paper #00911, Washington, DC

Hima H, Santibanez C, Roshan S and Lomme R (2016) The Nigeria Fadama National Development Series: how to build a pilot into a national program through learning and adaptation. Delivery case studies Doing Development Differently (DDD), The World Bank's Nigeria Country Team

HLPE (2017) Nutrition and food systems. A report by the High Level Panel of Experts on Food Security and Nutrition of the Committee on World Food Security

Hogendorn JS (1969) Economic initiative and African cash farming: pre-colonial origins and early colonial developments. In: Gann LH, Peter D (eds) Colonialism in Africa, 1870–1960: The economics of colonialism. Cambridge University Press, London

Hopkins AG (1973) An economic history of West Africa. Columbia University Press, New York

IFAD-IFPRI (2011) Agricultural commercialization in northern Ghana, innovative policies on increasing access to markets for high-value commodities and climate change mitigation. IFAD-IFPRI Partnership Newsletter, Washington, DC

IITA (1976) IITA annual report and research highlights, 1975

Jama B, Pizarro G (2008) Agriculture in Africa: strategies to improve and sustain smallholder production systems. Ann NY Acad Sci 1136:233–242. New York Academy of Sciences

Jayne TS, Ameyaw D (2016) Africa's emerging agricultural transformation – evidence, opportunities and challenges. In: Alliance for Green Revolution in Africa (AGRA). Africa agriculture status report: Progress towards agriculture transformation in sub-Saharan Africa. Issue No. 4, Nairobi, pp 1–20

John DA and Babu GR (2021) Lessons From the Aftermaths of Green Revolution on Food System and Health. Frontiers in Sustainable Food Systems 5:644559. https://doi.org/10.3389/fsufs.2021.644559

Kaufman F (2010) The food bubble: how Wall Street starved millions and got away with it. Harper's Magazine, 27–34 July

Khor M (2000) Globalization and the south: some critical issues. Third World Network. Spectrum Books Limited

Kijima Y, Otsuka K, Dick S (2011) An inquiry into constraints on a green revolution in sub-Saharan Africa: the case of NERICA rice in Uganda. World Dev 39(1):77–86

Kinyata GS and Abiodun NL (2020) The Impact of Community Participation on Projects' Success in Africa: A Bottom Up Approach. International Journal of Research in Sociology and Anthropology. Volume 6, Issue 3. https://doi.org/10.20431/2454-8677.0603001

Kolavalli S, Benin S, Babu S, Poku J, Thompson NM, Asenso-Okyere K (2008) Public expenditure and institutional review of Ministry of Food and Agriculture. IFPRI, Washington, DC

Lele U (1981) Rural Africa: modernization, equity, and long-term development. Science 211(6):547–553. (Also, World Bank Reprint Series: Number 186)

McIntyre BD, Herren HR, Wakhungu J, Watson RT (2009) Agriculture at a cross roads: international assessment of agriculture knowledge, science and technology for development – global report. International Assessment of Agricultural Knowledge, Science and Technology for Development (IAASTD). Island Press, Washington, DC

MOFA (2010) Medium term agriculture sector plan (METASIP) I, Accra

Mohan PC (2002) Nigeria: the National Fadama Development Project. Africa Region Findings & Good Practice Infobriefs No. 83. World Bank, Washington, DC

Morris M, Kelly VA, Kopicki R and Byerlee D (2007) Fertilizer Use in African Agriculture: Lessons Learned and Good Practice Guidelines. The World Bank, Washington, DC

Murgai R, Ali M, Byerlee D (2001) Productivity growth and sustainability in post-green revolution agriculture: the case of the Indian and Pakistan Punjabs. World Bank Res Obs 16(2):199–218

Ninan KN, Chandrashekar H (1992) The green revolution, dryland agriculture, and sustainability: insights from India. In: Peters GH, Stanton BF (eds) Sustainable agricultural development: the role of international cooperation. International Association of Agricultural Economists (IAAE), pp 102–114

Norman DW, Simmons EB, Hays HM (1982) Farming systems in the Nigerian savannas. Westview Press, Boulder

Oculi O (1981) Food imperialism and African diplomacy in the 1980s. Afr Dev 6(3):CODESRIA

Okigbo P (1985) Planning the Nigerian economy for less dependence on oil. Distinguished Lecture No. 3. In: Nigerian Institute of Social and Economic Research (NISER), Ibadan

Okorie A and Idachaba FS (1983) Farm Input Distribution Arrangements for Nigeria's Green Revolution Programme: Lessons from the World Bank-Assisted Projects. Project Report, Department of Agricultural Economics, University of Ibadan, Nigeria

Olayemi JK, Dittoh S (1995) Macroeconomic policies and the agricultural sector. In: Iwayemi A (ed) Macroeconomic policy issues in an open developing economy: A case study of Nigeria. National Centre for Economic Management and Administration (NCEMA), Ibadan, pp 441–464

Paitnak U (1972) The development of capitalism in agriculture. Social Scientist, September

Pattanayak CM (1983) Sorghum Breeding - Past, Present and Future: ICRISAT Cooperative Program in Upper Volta. Report presented at the Joint Meeting of the UNDP Joint Advisory Committee, ICRISAT, India

Pingali PL (2012) Green revolution: impacts, limits, and the path ahead. PNAS 109(31):12302–12308

Pretty J, Toulmin C, Williams S (2011) Sustainable intensification in African agriculture. Int J Agric Sustain 9(1):5–24

Ricciardia V, Navin Ramankuttya N, Mehrabia Z, Jarvisa L, Chookolingoa B (2018) How much of the world's food do smallholders produce? Glob Food Sec 17:64–72

Roessler P, Yannick I, Pengl YI, Marty R, Titlow KS, van de Walle N (2022) The cash crop revolution, colonialism and economic reorganization in Africa. World Dev 158:105934

Roling N, Jiggins J (2009) Making transdisciplinary science work for resource-poor farmers. In: Workshop farmer first revisited, Brighton, 12–14 December. Wageningen University, The Netherlands

Rosset PM (1999) Small is bountiful. Ecologist 29:2–7

Salau AT (1986) River basin planning as a strategy for rural development in Nigeria. J Rural Stud 2(4):321–335

Shepherd AW (1981) Capitalist agriculture in Africa. Afr Dev VI(3):1–7

Shiva V (1991) The green revolution in the Punjab. Ecologist 21(2):57–60

Stern E (1984) The evolving role of the bank (World Bank) in the 1980s. Paper presented at the agricultural symposium. The World Bank, Washington, DC

Stifel LD (1986) Director-general's report. In: IITA annual report and research highlights, 1985

Swaminathan MS (1983) Agricultural progress—key to third world prosperity. Third world lecture. South-South conference on strategies of development, negotiations and cooperation, in Beijing on 4 April, 1983

Swaminathan MS (2006) An evergreen revolution. Crop Sci 46(5):2293–2303. https://doi.org/10.2135/cropsci2006.9999

Toenniessen G, Adesina A, Devries J (2008) Building an alliance for a green revolution in Africa. Ann NY Acad Sci 1136:233–242. New York Academy of Sciences

Tosh J (1980) The cash-crop revolution in tropical Africa: agricultural reappraisal. Afr Aff 79(314):79–94

UNEP (2016) Food systems and natural resources. A report of the Working Group on Food Systems of the International Resource Panel

Vidal J (2011) Food speculation: 'people die from hunger while banks make a killing on food'. The Observer, 23 January

von Braun J, Afsana K, Fresco LO, Hassan M (2021) Food systems: seven priorities to end hunger and protect the planet. Nature 597(2):28–30

Weaving RV (1993) Agricultural development projects in Nigeria (English). OED Precis, No. 50. World Bank Group, Washington, DC

Welch RM and Graham RD (2000) A new paradigm for world agriculture: Productive, sustainable, nutritious, healthy food systems. Food Nutr. Bull. 21:361–366

WFP (World Food Programme) (2017) Food systems. http://www1.wfp.org/food-systems. 29/07/2017

World Bank (1975) The assault on world poverty. World Bank Press, Washington, DC

World Bank (1981) Accelerated development in Africa: an agenda for action, Washington, DC

World Bank (1995) Nigeria impact evaluation report: Kano Agricultural Development Project (Loan 1982–UNI), Sokoto Agricultural Development Project (Loan 2185-UNI). Report No. 14767-UNI, Operations Evaluation Department. World Bank, Washington, DC

World Bank (2007) World Development Report 2008: Agriculture for Development. The World Bank, Washington, DC.

World Bank (2009) Awakening African sleeping Giant: prospects for commercial agriculture in the Guinea Savannah Zone and beyond. World Bank, Washington, DC

World Bank (2017) Ghana: agriculture sector policy note – transforming agriculture for economic growth, job creation and food security. World Bank, Washington, DC

Integrating Indigenous and Modern Knowledge Systems for Household Food Security in the Smallholder Irrigation Schemes in South Africa

3

Sandile Phakathi and Sikhulumile Sinyolo

Abstract

There is currently a debate on the role of modern and indigenous knowledge systems in the smallholder farming sector. While the modern knowledge systems have been viewed as superior and are being touted as key towards meeting sustainable development goals, especially goal two of eradicating hunger, the performance of smallholder farmers utilising modern systems remains below expectations. The performance of farmers using indigenous knowledge systems has not resulted in better outcomes either. There are growing calls for an integration of the two knowledge systems to co-produce solutions relevant to smallholder farmers. However, not much is known about the extent to which this is already happening, how the integration is happening, and welfare effects of this integration. This paper assessed the extent to which smallholder irrigators integrate indigenous and modern knowledge; and whether this integration has resulted in improvements in cost reduction,

productivity, and welfare levels. A sample size of 392 farmers in four irrigation schemes of KwaZulu-Natal and Eastern Cape were interviewed. The results showed that the majority (52%) of the farmers integrated knowledge systems, while 10% and 38% relied on indigenous and modern systems, respectively. Farmers actively incorporated modern knowledge to enhance their traditional practices in their production methods in response to local challenges or opportunities. The results indicate that high maize yields attained by modern users were eroded by high inputs cost, while integrators were able to significantly reduce input costs (such as fertilizer, pesticide, herbicide costs) and the money saved was used to purchase more food, hence improving household food security. The results suggest that government, private institutions and NGOs should build on farmers' agencies to plug-in modern knowledge into the indigenous knowledge systems through appropriate research and innovations rather than completely replacing indigenous systems with modern farming systems. Extension officers should change their mindset and acknowledge the importance of integrating indigenous knowledge into their training so that what is relevant to farmers may be adopted, thus enhancing sustainability and resilience of development efforts in developing countries.

S. Phakathi (✉)
Rhodes University, Makhanda, South Africa

S. Sinyolo
Human Sciences Research Council,
Pretoria, South Africa

© The Author(s) 2025
S. Dittoh et al. (eds.), *Integrating Indigenous and Scientific Knowledge for Sustainable Food Systems in Africa*, Sustainable Development Goals Series,
https://doi.org/10.1007/978-3-031-85512-2_3

29

Keywords

Knowledge systems · Sustainable development goals · Plug-in principle · Sustainable livelihood framework · Smallholder irrigation · South Africa

3.1   Introduction

Smallholder farming plays an important role in the livelihoods of the poor rural households in developing countries. Wide consensus exist that smallholder farming should be prioritised to achieve the Sustainable Development Goals (SDGs) by 2030, given its accessibility to the marginalised (Ndabeni 2019). However, these poor smallholder farming households generally experience low crop productivity levels, and oftentimes, total crop failure, leading to poverty and increased vulnerability to food insecurity (Hendriks 2013).

There are debates on the reasons why smallholder farming remains trapped in underproductivity, particularly on the role of indigenous and modern systems. On one side, several authors have argued that the low crop yields are due to limited adoption of modern and improved inputs and technologies (Gatzweiler and Von Braun 2016; Triomphe et al. 2013). While these studies have largely shown that the adoption of modern technologies are associated with increased yields, they have also noted that the cost of these technological modern systems are high, which has meant that smallholder farmers, who are the liquidity-constrained, do not afford them (ACB 2012; Sinyolo 2020). Others have argued that the limited adoption of modern systems is because the success of these systems is not certain, as most succeed under stringent managerial regimes and agro-climatic conditions which are beyond the reach of the smallholder farmers (Gatzweiler and Von Braun 2016; Gizaw 2003). Additionally, limited market access, inadequate storage and transport infrastructure, as well as increased chances of buying counterfeits, reduce the incentive of smallholder farmers to invest in the modern agricultural technologies (Sinyolo 2020).

On the other hand, some studies have argued that the focus on modern scientific solutions, while paying insufficient focus on indigenous systems, which rely on traditional and local knowledge to develop new and improved farming methods, is to blame (Buthelezi and Hughes 2014; Hart and Mouton 2005; Ncube et al. 2018). While indigenous knowledge has been defined in various ways by different authors, these definitions highlight the idea that it is 'a cumulative body of knowledge, practice and belief handed down through generations by cultural transmission'(Gómez-Baggethun   2022). Scientific knowledge has been privileged, and indigenous knowledge suppressed (Briggs and Moyo 2012). Reliance on indigenous knowledge has been promoted in recent decades by many as an alternative way of achieving development among poor smallholders in many developing countries (Briggs 2005; Briggs and Moyo 2012). Indigenous knowledge is not only accessible but has also been promoted in latest debates about climate resilience and sustainable development because traditional practices have allowed people to live in harmony with nature (Briggs 2005). As highlighted in Briggs and Moyo (2012), it is the cultural and economic embeddedness of indigenous knowledge that ensures its resilience across Africa, as the farmers use, develop, rework and rely on the locally embedded knowledge which they have by necessity developed for themselves for generations.

A focus of the growing literature on traditional knowledge systems has shown the extent of reliance on indigenous knowledge, with several case studies documenting these traditional practices been developed (Apraku et al. 2022; Buthelezi et al. 2013; Gómez-Baggethun 2022; Hart and Mouton 2005; Nyong et al. 2007; Ubisi et al. 2019). These argue that while smallholder farmers are producing viable solutions, these go either unnoticed, or neglected as being outdated, inefficient and unproductive (de Bont et al. 2019). These case studies reveal how local farmers have developed innovative ways of survival using indigenous knowledge, particularly in the context of climate change (Buthelezi 2018). For example, Apraku et al. (2022) found that smallholder farmers in South Africa and Kenya relied on a wide range of traditional or indigenous agricultural practices, customs, belief systems and skills to boost their agricultural activities in the face of changing global and local climatic conditions. The study found that indigenous knowledge was used in predicting seasonal weather and rainfall patterns, determining wind speed and direction, preserving grains for planting purposes and various traditional farming support systems to lessen the impacts of climate change on their agricultural activities. However, this focus on indigenous knowledge has not resulted in expected outcomes, with those that rely on it remaining stuck in low productivity (Briggs and Moyo 2012). The future of indigenous knowledge as a development tool is doubtful, and Briggs and Moyo (2012) argue that it is more likely to be permanently located at the margins of development practice.

There is a growing literature (Agrawal 1995, 2009; Basdew et al. 2017; Briggs and Moyo 2012; Retnowati et al. 2014; Bolosha et al. 2023; El-Hani et al. 2022; Hiwasaki et al. 2014; Malone et al. 2022; Hermans et al. 2022; Ray 2023; Wang 2015) that has argued that the binary debates of modern versus traditional, or scientific versus indigenous knowledge are futile. A focus on the distinctions between indigenous and scientific knowledge risk the exclusion of diverse forms of expertise and solutions relevant for local contexts (El-Hani et al. 2022). These studies have argued for an integration of the two knowledge systems to co-produce solutions that are informed by the needs, experiences, capacity, and desires of smallholder farmers. Such integration is crucial for developing effective local adaptation strategies and actions. The argument is that smallholder farmers are not just adopters and users of modern systems, but that they integrate modern knowledge into their indigenous knowledge system (the Plug-in Principle), making important incremental changes on modern innovations adopted from elsewhere to suit their contexts and demands (Bolosha et al. 2023; Dittoh 2003). As explained in the Chap. 1 of this book (Dittoh), the Plug-in Principle argues that achieving greater, more effective, and sustainable integration

requires that scientific or modern knowledge is "plugged in" to traditional practices, since it will be plugged into what farmers already know. It acknowledges the farmers as possessing important knowledge about their environment, and that greater success can be achieved when there is integration between modern and traditional knowledge, not replacement of traditional knowledge. As Clemens (2021) argued, innovations are a result of an encounter of diverse knowledges at the local level.

Literature that has documented examples of integration of modern and indigenous knowledge systems (Yanou et al. 2023), have mainly focused on the importance of indigenous knowledge systems. Another strand in the literature have focused on debates on the best or appropriate ways of merging indigenous and modern knowledge to produce more effective hybrid ways of knowing (Briggs and Moyo 2012). These studies have generally been descriptive and place-specific; and have rarely sought to investigate the impacts of this integration. The objective of the study is to assess the extent to which smallholder irrigators integrate indigenous and modern knowledge; and whether this integration has resulted in improvements in outcomes such as cost reduction, productivity, and welfare levels.

3.2 A Systems Approach to Promoting Innovation in the Smallholder Farming Sector

The pipeline approach to agricultural technology innovations remains an influential approach for understanding innovation in the farming sector. This linear model of innovation postulates that innovation begins with basic research, and that researchers develop modern technologies which they pass to extension agents for dissemination to the farmers (Triomphe et al. 2013; Godin 2006). However, despite significant investments for Sub Saharan Africa, this approach has not led to satisfactory returns, as the farmers, who are the end-users of innovations, are considered passive adopters (Malek et al. 2017; Pamuk et al. 2014).

The linear model has thus been abandoned, and the innovation systems approach, which is more participatory, inclusive and holistic, promoted (Malek et al. 2017; Kebebe 2019). The national system of innovation approach is the preferred model of promoting innovation in South Africa, as entrenched in the recent White Paper on Science, Technology and Innovation (DSI 2019).

The innovation systems model indicates that innovation is not linear, but occurs within heterogeneous networks, characterised by a diversity of stakeholders, which include government, researchers, farmers, private entrepreneurs and non-governmental organizations (NGOs) (Triomphe et al. 2013). These innovation actors interact in a non-linear, iterative and non-predictable pattern to solve a common problem, adapt to a new environment or take advantage of new opportunities (Triomphe et al. 2013). The smallholder farmers are not only confined to being users of innovation, but also as producers of innovations (Lundvall 2009; Von Hippel 2005; Gupta 2016; Heeks et al. 2014). The growing literature on inclusive innovation (e.g.,Cozzens and Sutz 2014; Heeks et al. 2014; Santiago 2014) emphasizes the importance of not only producing innovations that benefit the disenfranchised or marginalised communities, but the participation of these marginalised in the innovation processes. This requires consideration of the marginalised actors such as smallholders as beyond users, but as generators of innovations. Smallholder farmers rely on a wide range of agricultural knowledge sources to inform their own production systems, which includes their own accumulated experiences, their own experimentation and from scientific and agronomic information available (Briggs and Moyo 2012). According to Spielman et al. (2009), innovation can only have socioeconomic impact when it is part of sustained processes involving many actors with different capabilities and resources. While the innovation system concepts is entrenched in the broader literature, they remain poorly understood in the agriculture content, with little empirical documentation (Triomphe et al. 2013; Spielman et al. 2009; Malek et al. 2017). Both modern and indigenous knowledge are acknowledged, however,

modern knowledge is considered superior, with indigenous knowledge expected to increasingly play a reduced role as it gets replaced by modern knowledge (Dittoh 2003; Hart and Mouton 2005). The Plug-in Principle challenges this thinking, highlighting a central role of the indigenous knowledge system, with modern knowledge getting integrated in a way that strengthens the local knowledge system.

3.3 Factors Influencing Decision to Adopt a Particular Farming Practice

Household socio-economic characteristics (such as age, education, gender, land size, access to financial credit, access to extension officers) influence farmers' decision to rely on a particular innovation knowledge system. According to Adesina et al. (2000), older farmers are more likely to use integrated knowledge systems because, with age, farmers get exposed to various farming challenges and opportunities, and they are more likely to experiment and integrate new farming practices. On the other hand, (Amsalu and de Graaff (2007), argued that younger farmers are more likely to adopt modern innovation systems because they are more flexible and can search for new information to explore new farming practices.

There is a positive relationship between education and adoption of modern innovation systems (Oduro Ofori et al. 2015). Okpachu et al. (2014), argue that education is imperative in advancing technologies as farmers must understand complex scientific changes to adapt and remain productive. The effect of education on farmers is two-fold, cognitive and non-cognitive, with the cognitive effects consisting of basic literacy and numeracy that farmers gain from education (Oduro Ofori et al. 2015). Literacy enables farmers to read and comprehend information; for example, on inputs such as fertilizers while numeracy enables farmers to calculate the sufficient measurements needed to get the required output (Ninh 2020). However, education plays a larger role for early inventors who use their education to reduce the costs of acquiring information and use it to learn new methods, while the other farmers merely adopt these practices (Weir and Knight 2014). This gives farmers that are educated an upper hand as they are able to make calculated decisions. The majority of studies on smallholder agriculture in South Africa note that farmers' education levels are relatively low, with the majority never having attended school (Cele and Wale 2018; Dirwai et al. 2019; Muchara et al. 2016). Therefore, educated farmers are expected to rely on modern systems than indigenous farming practices because they can better process information and seek appropriate technologies to address production constraints.

Empirical studies have found that male farmers are more likely to rely on modern farming systems (Shiferaw et al. 2011b; Mariano et al. 2012). This is because they have better access to capital assets, such as financial assets and larger plot sizes required to efficiently use modern or integrated innovation systems (Shiferaw, Bank, et al. 2011a). Farmers with large farm sizes are more likely to adopt modern innovation knowledge systems because they tend to use intensive farming practices, and returns from economies of scale are better when cultivated land is larger (Ghimire and Huang 2016). Household labour size has been found by Suvedi et al. (2017) to have a positive influence on the adoption of modern knowledge systems because it signifies the availability of labour required for intensive practices as compared to indigenous farming systems.

Empirical studies have found that access to extension services positively influences reliance on modern farming systems as compared to indigenous knowledge (Mignouna et al. 2011; Raut et al. 2011; Asfaw et al. 2012; Mariano et al. 2012; Arslan et al. 2014). A functioning extension service system is imperative for information dissemination and influencing the adoption of modern farming systems in rural areas as it is still viewed as superior and better than indigenous systems (Ghimire and Huang 2015). Therefore, extension officers play a crucial role in changing farmers' attitudes and behaviours on the adoption and practices of modern innovation systems,

thereby decreasing the knowledge asymmetry that new technologies are frequently connected with (Ghimire and Huang 2015). On the other hand, extension officers have been critiqued for being too narrow when providing their services, as they tend to provide training that is not responsive to farmers' needs (Chowa et al. 2013). This requires mutual learning from extension services and farmers so that the bettering process and integration of knowledge systems can be improved.

Moreover, participation in any collective action significantly influences farmers' decisions on the type of knowledge system to practice. Conley and Udry (2010) noted that participating in farmer groups enables farmers to learn how to grow new crops, modify and integrate farming practices to suit local context. Several studies (eg, Mignouna et al. 2011; Sinyolo and Mudhara 2018) found a positive association between participating in a farmer group and the adoption of modern innovation systems. Farmers in a social network get exposed to various knowledge sources, weigh the cost and benefits and take advantage of new knowledge to create integrated practices to solve their production challenges.

3.4 Methodology

3.4.1 Sampling Procedure and Study Area Description

The Eastern Cape and KwaZulu-Natal were selected as the study areas because smallholder farming constitutes a key livelihood option for poor rural farmers. Irrigation farming is crucial in these areas because of high temperatures and low precipitation rates, which makes dry land farming a risky livelihood venture (Muchara et al. 2014). The study employed a multi-stage and purposive sampling strategy to select participants (*see* Phakathi et al. 2021 *for further details*). Firstly, Eastern Cape and KwaZulu-Natal were purposively selected because of the importance of smallholder irrigation. Secondly, four irrigation schemes were purposively selected from a list of irrigation schemes in the Eastern Cape and KwaZulu-Natal provinces. The lists of irrigation

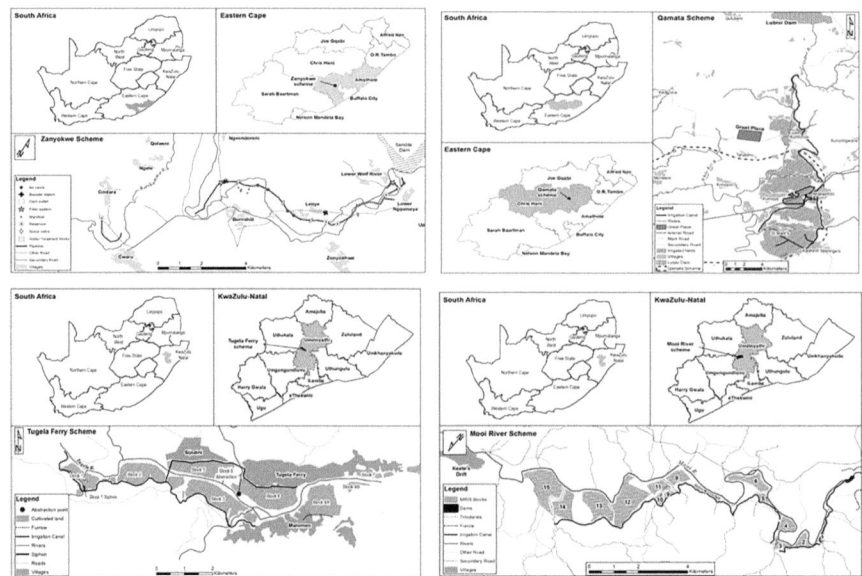

Fig. 3.1 Study area

Table 3.1 Key characteristics of the selected irrigation schemes

	Tugela Ferry	Mooi-River	Qamata	Zanyokwe	Average
Province	KZN	KZN	EC	EC	
Year formed	1989	1902	1960	1983	
Duration (years)	29	116	58	35	60
Irrigable Size (ha)	837	600	400	450	572
Main canal length (m)	34	25	28	–	29
Average plot size (ha)	0.2	0.4	2	3	5.6
No. of irrigators	1500	824	350	164	710
No. of blocks	7	15	7	6	9
No. of blocks selected	7	9	6	6	7

Source: Phakathi et al. (2021)

schemes were obtained from the provincial departments of agriculture. After field visits to most of the schemes and informal discussions with farmers on scheme dynamics, the Qamata and Zanyokwe irrigation schemes of the Eastern Cape and Tugela and Mooi River irrigation schemes of KwaZulu-Natal were selected. Figure 3.1 shows the location of the four schemes. A sample of 392 irrigation farmers was selected and interviewed. The extension officers provided the list of farmers participating in the irrigation schemes and assisted with setting up individual

interviews and focus group discussions. The aim was to assess who is more likely to integrate or mainly primarily rely on indigenous or modern systems given different contexts. Pooling participants from different geographic regions enabled comparison.

Table 3.1 presents the key characteristics of the four selected irrigation schemes. The Tugela Ferry irrigation scheme was established in 1989, while the Mooi River in 1902 in KwaZulu-Natal. The Zanyokwe scheme and Qamata in the Eastern Cape were formed in 1983 and 1960, respectively

(Phakathi et al. 2021). These four schemes have been operating for over 60 years on average. The total land size irrigated in hectares varies across schemes, with 837 in Tugela, 600 in Mooi-River, 400 in Qamata and 450 in Zanyokwe (Phakathi et al. 2021). The average plot size allocated per farmer ranges from 0.2 to 3 hectares (0.2 in Tugela, 0.4 in Mooi-River, 2 in Qamata and 3 in Zanyokwe on average. Moreover, the Zanyokwe has fewer irrigators (164), followed by Qamata at 350. Tugela ferry has many beneficiaries' of 1500 irrigators while 824 are located in Mooi-River. To manage water resources, farmers are categorised into smaller manageable blocks sharing the shame secondary irrigation systems.

Data were collected in 2019 by four enumerators in each scheme (16 in total) who spoke IsiZulu and IsiXhosa, the home languages in each scheme using structured questionnaires and focus group discussions. The questionnaires were translated into the farmers' home languages. The enumerators were trained, and the questionnaires were pretested before the survey.

3.4.2 Process for Categorising Farmers as Indigenous, Modern and Integrators of Knowledge Systems

Innovations can result from traditional or indigenous knowledge. Innovations produced through traditional knowledge are less costly and are often suited to farmers' management experience and environment contexts. However, innovations based on traditional knowledge systems have been found to be less productive. The second type of innovations is the adoption and use of modern knowledge systems. While potentially more productive, these innovations are often too expensive, or less suited to the experience and environmental contexts of the farmers. It is in this context where an intermediate form of innovation is promoted, that which relies on the integration of modern and indigenous knowledge. In this integration, there are three potential approaches.

First, a traditional production practice can be enhanced. That is, scientific or modern knowledge is used to improve traditional practices, described as "plug-in" in this book. Second, a modern innovation can be adapted to suit the local context. In this case, the innovation that was developed using modern knowledge is adapted to the local context by making adjustments informed by indigenous knowledge, a reverse "plug-in", where indigenous systems are plugged into modern systems. The third, which is more radical, involves the integration of both traditional and modern knowledge systems to produce a new or improved production practice that is completely different. In this paper, we included all of these three types.

To identify whether a farmer relies heavily on indigenous, modern or integrates knowledge systems, they were asked to identify any type of innovations or knowledge systems implemented in the past three years around farming methods, soil and water conservation measures, fertiliser use, herbicide use, irrigation management and storage facilities. Based on the reported innovations or knowledge systems, farmers were categorized into three groups, as (1 = indigenous, 2 = modern, and 3 = integrators) user of knowledge systems. In this paper, innovations were not considered as something new to the community but as a new practice implemented by a farmer regardless of whether it already exist somewhere. The period of three years was considered long enough to capture innovation activities and reduce recall problem (Armbruster et al. 2008). Extension officers and other development officials participated in some of focus group discussions and assisted in giving further information on some of the innovations implemented by the irrigators over the years, as they work closely with farmers. The study also collected data on socio-economic characteristics and food security indicators (food consumption expenditure, household food insecurity hunger scale, household hunger scale, and income level) to identify the differential characteristics among these farmers based on the innovation knowledge systems used.

3.5   Results and Discussion

3.5.1   Extent to Which Smallholder Irrigators Integrate Indigenous and Modern knowledge

Figure 3.2 below present the percentage of farmers based on the knowledge system they utilised in implementing farming practices. The results show that out of a sample size of 392, 10% of farmers were reliant on indigenous knowledge systems only, while 38% were dependent on modern farming systems only. Most of the farmers (52%) integrated both modern and traditional knowledge systems in their production activities. While the dependency on traditional knowledge systems is limited, the results show that farmers prefer to mix both modern and indigenous farming systems. According to the farmers, the integration of both modern and indigenous knowledge systems was motivated by the desire to manage costs, on one hand, while benefiting from potential yield increases, on the other. For others, the main motivation was to adapt the modern practices and technologies to their local contexts. The most popular type of integration was making improvements or adjustments on the adopted technologies so that they suit the farmer's context and needs.

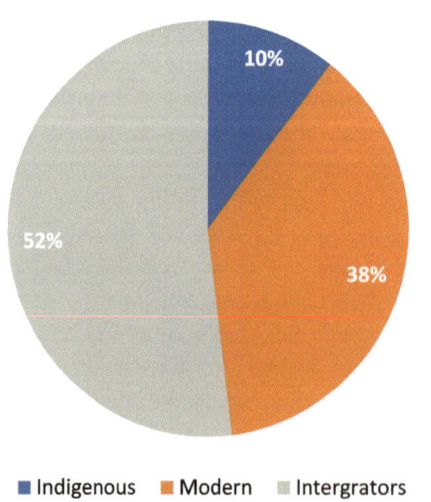

Fig. 3.2 Reliance of farmers on different knowledge systems

Discussions with the farmers indicated that they relied on different knowledge systems, practices and technologies for the various land preparation, planting, weed and pest control, and harvesting activities. Farmers integrated their indigenous knowledge system to supplement modern practices. For instance, they integrated the indigenous method of gauging seedling depth through manual measurement with their fingers, alongside adopting the modern technique of adhering to the 4 cm depth rule for determining seedling depth during planting, a practice conveyed by extension officers during training sessions.

Moreover, in response to the challenge of high seedling costs, farmers adopted an integrated approach. Some plots were sown with hybrid seeds, while others were cultivated using traditional landraces or recycled seeds. Regarding land preparation, farmers employed a blend of modern and indigenous methods. They utilised traditional practices such as employing oxen for land preparation to reduce production costs, alongside employing tractors when financially feasible. Moreover, farmers diversified their fertilisation strategies beyond reliance solely on modern fertilisers. They incorporated organic fertilisers, such as cow manure, and employed ashes to combat pests. The application of ashes involved mixing them with water and applying the mixture using a broom. To standardize application, farmers utilised the lid of a Smirnoff container for each vegetable plant. Additionally, they supplemented these homemade mixtures with purchases of modern fertilizers to fulfil the necessary inputs for optimal plant growth.

Furthermore, farmers indicated that storage facilities are one of the major bottlenecks hindering sustainability of their farming livelihood. The majority of the farmers in these schemes do not have proper storage facilities such as cold rooms, barns, granaries and storerooms to store their produce after harvesting. This is a significant bottleneck as produce perishes, and farmers cannot sell during the off-season to benefit from higher prices. To mitigate this issue, some farmers still resort to traditional storage methods, including utilising cribs for maize and employing river sand to create a cool, dry environment for storing vegetables. While on the other hand, some farmers have innovatively repurposed sections of their household roofs, transforming them into grass-covered storage spaces, as depicted in Fig. 3.3. This adaptation enables them to maintain produce freshness for extended durations in a cool, dry setting. Additionally, some farmers opt to dig holes outside their houses to store potato seedlings intended for planting in subsequent seasons.

According to the sustainable livelihood framework, farmers cannot reap all the potential benefits without the equipment required to make farming a success (Ellis et al. 2003; Scoones 2009). The government and private donors can significantly assist farmers, especially with storage facilities, which are important for storing perishable agricultural products to make farming a sustainable livelihood but not replace the use of indigenous practices.

3.5.2 Costs and Yield Differences by Knowledge System

To analyse the costs and yield differences, we focused on maize, which was produced by most of the farmers in sample. Table 3.2 presents costs and yields by type of knowledge systems maize producers relied on. The results in the table show that indigenous users' inputs are relatively cheaper as compared to the other two in terms of fertiliser, herbicide, and pesticide costs. The farmers who relied on modern inputs only incurred the highest costs, while those who merged modern and indigenous inputs incurred costs that were between the two extremes.

Table 3.2 shows that those who use modern knowledge systems achieved higher yields than the other two categories. Those who relied on indigenous knowledge systems achieved the lowest yield per hectare, almost 3 times lower than that of modern based production systems. This is expected and is the main reason why modern technologies are promoted as a key strategy to achieve food security among smallholder farming households. However, it is not enough to compare yields, as the costs of these production systems are not the same. Even though other factors have not been controlled for, the profit figures based on input cost indicate that, while yield differences are very huge, the net profits are not very different. This is because the yields achieved by those who rely on modern inputs come at a higher cost. Even though the government assists sometimes through purchasing inputs for them, farmers indicated that these inputs often arrive very late, or the government buy incorrect inputs which are not good for their soils or crops they had planned to cultivate. This results in farmers planting late in the season, leading to low productivity (Phakathi and Wale 2018). Integration of knowledge system should

Fig. 3.3 Example of integrated storage facility used by a farmer in Zanyokwe scheme

Table 3.2 Average maize production inputs costs and yield across farmer typologies

Maize inputs per hectare	Means = 392			F-test
	Indigenous (n = 41)	Modern (n = 147)	Integrators (N = 204)	
Fertilizer cost (Rands/kg)	34.78	572.93	404.95	***
Herbicide cost (Rands/litre)	11.73	224.43	148.14	***
Pesticide cost (Rands/litre)	30.43	270.94	162.18	**
Maize yield (kg/ha)	403.04	1198.98	929.98	***

Source: Field Survey (2019)

Table 3.3 Socio-Economic variables among farming typologies

Variables	Means = 392			F/χ^2-tests
	Indigenous (n = 41)	Modern (n = 147)	Integrators (N = 204)	
Age	54.01	51.37	55.28	***
Education	4.22	7.14	2.40	**
Gender (1 = female)	0.66	0.53	0.79	***
Farming experience	20.44	15.90	21.15	**
Household size	6.24	6.00	6.4	
Leadership scheme (chair, treasury, secretary)	0.05	0.29	0.26	***
Land size in hectares	0.9	1.8	1.6	**
Water access (no of days)	2.85	4.78	4.13	**
Assets value (logged)	8.35	8.70	8.70	
Livestock TLC	1.80	1.84	1.97	

Source: Field Survey (2019)

be fully exploited as farmers can benefit more from using both indigenous and modern practices to suit their local context.

3.5.3 Differential Characteristics Based on Socio-Economic Status

Table 3.3 presents the results on the association between socio-economic variables and farming typology based on the knowledge system used by a farmer. The results show that older farmers dominate smallholder irrigation schemes with an average age of more than 51 years. It remains a challenge to attract youth to the agricultural sector as farming is regarded as unattractive because it takes months to generate an income. The significant p-value suggests that the mean age differences among the indigenous, modern and integrators of knowledge systems are significant at the 1% significance level. The results show that older farmers are more likely to use indigenous and integrated practices in their production, and less likely to use modern practices. The finding is similar to other studies which found that older farmers are not likely to be the users of modern innovation systems because they are not either physically or mentally strong in taking more efforts and risks associated with the adoption of modern knowledge systems (Kassie et al. 2011; Asfaw et al. 2012). In contrast, younger farmers are still flexible and willing to take more risks.

The results show relatively low levels of education among these farmers, with an average level of 7 years of education completed among those who rely on modern systems, 4 years of education completed among those who rely on indigenous systems, and about 2 years among those who integrate the two knowledge systems. This indicates a relatively low level of the human capital among the farmers. The significant p-value suggests that the mean education level differences among the indigenous, modern and integrators of knowledge systems are significant at the 5% significance level. The results show that modern farmers are more educated compared to other groups, with an average education level of grade seven. Modern knowledge systems require farmers to be able to acquire, interpret and understand the information disseminated on modern systems. Formally educated farmers are able to take the initiative and create innovations by introducing new ideas as found by other studies (Okpachu et al. 2014; Weir and Knight 2014). An empirical study that was conducted by the World Bank in 1992, found that farmers with basic education were 8.7% more productive than farmers without the basic education (Oduro Ofori et al. 2015). An educated farmer is more likely to make better choices and make economically rational choices (Oduro Ofori et al. 2015). Education also allows farmers to be able to learn more easily through observation from other works of innovation than those who are not, they are also able to weigh the costs and advantages to help determine whether or not to adopt, which allows farmers to adapt to inventions which best suit their context efficiently (Weir and Knight 2014). An interesting result is that farmers who rely on traditional knowledge systems were more educated than those who merged both knowledge systems. While both these groups have very low levels of education, the result shows that traditional knowledge systems is not just about lower education levels.

The results show that, on average, smallholder irrigation is dominated by female farmers, with 66% using indigenous practices while 79% integrates both modern and indigenous farming practices within their local context. Female farmers dominate because they contribute the most labour in the smallholder sector as males tend to migrate to urban areas while women remain in rural areas to care for their children (Cousins 2013). The significant p-value suggests that gendered differences of the levels of reliance on the indigenous, modern and integrators of knowledge systems. Females are less likely to rely on modern knowledge systems. Evidence from other studies have shown that women are less likely to adopt modern agricultural innovations due to gender inequalities in access to resources, assets and services (e.g., financial services) (Diiro et al. 2018; Shiferaw et al. 2011b). Given the high costs associated with modern inputs, female farmers are less likely to afford them. Therefore, unequal access to resources because of patriarchal systems create a multitude of socio-cultural obstacles for females and inhibits them from adopting modern innovations (Sinyolo and Mudhara 2018). Further, the gender roles in society where women need to be home and perform other works such as taking care of the children, cooking and cleaning leave women with little or no time to be associated in a farmer group or spaces where they can exchange information and resources and become innovative or continue to practice some of the innovations implemented (Ajewole et al. 2015).

The significant p-value suggests that the mean experience level differences among the indigenous, modern and integrators of knowledge systems are significant at the 5% significance level. The results show that farmers who uses indigenous and integrate practices have more experience compared to modern users of technology. Integrating knowledge systems require more trial-and-error experiences as farmers get time to test what's works. Moreover, experienced farmers acquire sufficient capacity to deal with various farming challenges over time (Ainembabazi and Mugisha 2014).

Mean land size significantly differed across the three categories of the farmers. The results show that farmers with bigger land size were more likely to adopt modern farming practices, while those with smaller land sizes were more likely to depend on indigenous innovations systems. The results are comparable to past empirical studies (Mariano et al. 2012; Shiferaw et al. 2011b). Farmers with large farm sizes adopt modern knowledge systems because they tend to use intensive farming practices and returns from economies of scale are better when land cultivated is larger (Shiferaw et al. 2011b; Mariano et al. 2012). Also, land size proxies wealth, and indicates that wealthier farmers can afford the more expensive modern inputs. Furthermore, the results show that farmers who are leaders in the scheme are the users of modern and integrated innovation systems.

3.5.4 The Role of Social Capital on Innovation Practices

The study further investigated the role of social capital on influencing the knowledge system practices used by farmers. The results across groups in Table 3.4 show a statistically significant difference when a farmer participates in a social group, such as *Stokvel* (informal savings society, in which farmers contribute an agreed amount and receive a lump sum at the end of the year), water use association, religious group and burial societies. The results signify the importance of being in a networking space in order to benefit from knowledge sharing. For example, *stokvels* lend money to their members at lower interest rate compared to non-members. Farmers were then able to get access to informal loans, which helped them cover some cost of productions. Moreover, relevant announcements on production or marketing information are made in burial societies.

Social capital reduces the costs that an individual farmer would have incurred in devising a solution to a particular challenge as it enables communities to overcome their socially faced problems through the integration of knowledge (Van Rijn et al. 2012). The International Fund for Agricultural Development, International Federation Fund for Agricultural Procedures, and other international organisations have shown that the involvement of farmer groups and their capability to deliver efficient demonstrations and services for small farmers is crucial in accomplishing sound rural development (Adekunle and Fatunbi 2012).

3.5.5 The Importance of Support Services on Farming Practices

The results in Table 3.5 show that indigenous knowledge users are located far from extension support services compared to those who use modern knowledge, and integrators of knowledge systems. Access to extension services has been found to be positively associated with the adoption of modern knowledge. Extension officers generally operate in the old mode of the linear model of innovation, acting as agents that transfer knowledge and technologies from agricultural research to farmers during training sessions. Therefore, farmers with more access to extension services incur lower information costs and are more likely to adopt modern innovation systems (Gebre and Zegeye 2014). In addition, extension services can help farmers gain resilience to shocks, as they can connect farmers with relevant private sector actors who are more knowledgeable about specific shocks (such as pest outbreaks) based on their expertise (Davis and Laas 2014). However, extension services tend to disseminate information using a top-down approach, which does not fit the local context and does not cater to the actual needs of the farmers in that environment (Davis and Laas 2014). This continues to perpetuate the ideas which perceive modern innovation systems as superior to indigenous systems. Integration of the knowledge system is crucial so that researchers develop technologies that are locally relevant to farmers' needs.

There is a statistically significant difference in terms of access to financial credit among the three farming categories. Farmers who use

Table 3.4 Social Capital Variable across farming typologies

Variables	Means = 392			
	Indigenous (n = 41)	Modern (n = 147)	Integrators (N = 204)	F-test
Farmer group	0.46	0.58	0.58	
Stokvel	0.12	0.72	0.62	***
Water use association	0.24	0.44	0.49	*
Religious group	0.80	0.45	0.63	**
Burial societies	0.32	0.21	0.36	***

Table 3.5 Access to support services

Variables	Means = 392			F-test
	Indigenous (n = 41)	Modern (n = 147)	Integrators (N = 204)	
Distance to extension service	14.66	13.6	13.91	
Access to credit	0.12	0.28	0.28	***

Table 3.6 Outcome variables across farmer typologies

Variables	Means = 392			F-test
	Indigenous (n = 41)	Modern (n = 147)	Integrators (N = 204)	
Income per annum	7673.10	21195.45	26671.82	**
Food expenditure (per month)	440.49	1345.03	1755.63	**
Food security poverty status	0.24	0.51	0.62	***
HFIAS food security status	0.34	0.55	0.71	***

mainly modern knowledge systems, and those who merge modern and traditional knowledge systems, had higher levels of access to credit than indigenous knowledge users. The results are expected because farmers facing liquidity constraints opt for the cheaper indigenous practices. This results in low take-up of modern or integrated farming technologies such as fertiliser, pesticides and improved seed varieties, directly reducing productivity (Morris 2007; Selejio and Lasway 2019). Access to credit is crucial because it improves farmers' liquidity and enables them to respond to adverse shocks given intensifying climate change.

3.5.6 Food Security Levels According to Knowledge System

Table 3.6 displays the households' levels of food security, as proxied by the household food insecurity access scale (HFIAS), and food expenditure levels. The table indicates that the farmers who integrate knowledge systems had higher incomes and consume more than the indigenous and modern knowledge users, indicating that integrators are more food secure than the indigenous and modern system users. Integrators managed to reduce input costs, which enabled them to use savings to improve food consumption intake, ceteris paribus.

To further establish the linkage between knowledge systems practices and food security,

the households were categorised into two classes according to their food consumption levels- food secure and insecure. Based on the minimum per capita food poverty line for 2019 (Stats SA 2019), a figure of R561, R810 for the lower bound poverty line and R1227 for upper bound line per capita month were used as the food security line. This figure was taken from that suggested by (Stats SA 2019) and then adjusted using the consumer price index (CPI) so that it reflected the 2019 purchasing power of the Rand. The households with food consumption level greater than the cut-off line were considered food secure, while those below were considered food insecure.

The results indicate a statistically significant association between the knowledge systems used and the food security status of the farmers. While the bigger proportion of both the indigenous and modern knowledge users was classified as food insecure, the same does not apply to farmers who integrate systems. The majority (62%) of the integrators were classified as food secure, further highlighting the importance of integrating both knowledge systems for sustainable food security in smallholder agriculture. This is because integrators saved more on input costs and achieved better yields than those who rely on indigenous knowledge systems. The majority of farmers who rely only on the indigenous knowledge system are considered poor and food insecure, as only 24% were above the food poverty line, while a total of 51% of modern users were categorised as food secure. The results are comparable also when using HFIAS indicators, with more farmers categorised as food secure across the group. This is because farmers also consume some of their produce during the season to boost their food security level. The results indicate that when used independently, indigenous innovation system is not optimal in the context of irrigation schemes but are optimal when integrated with modern knowledge systems.

3.6   Conclusions

Achieving sustainable development goals, especially eradicating poverty in Africa, requires a mindset shift from the binary debates of modern versus traditional towards an integration of the two knowledge systems. Modern knowledge systems, while preferred and considered superior to indigenous knowledge systems, has not achieved expected results. Smallholder farmers still rely on indigenous knowledge systems in their production activities. The results of this study have shown that most farmers integrate indigenous and modern systems into their practices, with more than half of the respondents reporting that they merged both modern and traditional knowledge systems. Farmers indicated that they engage in incremental innovations, plugging in modern and indigenous practices to improve or better their production activities according to their local challenges or opportunities. There were significant differences in the socio-economic characteristics of those who rely on indigenous, modern and integrators of knowledge systems, with variables such as age, education level, gender, extension support, access to credit playing a crucial role in these differences.

The results show that even though the use of modern knowledge systems led to improved yields, the high cost of modern inputs erodes the potential profits and welfare outcomes. On the other hand, indigenous knowledge system was associated with lower productivity and lower costs, which also did not augur well for food security. Integration of both modern and traditional knowledge systems was associated with better food security levels than either indigenous

or modern knowledge system on its own. These results suggest that policy makers should devise policies that favour an integration of modern knowledge into the traditional knowledge system, not those that favour one or the other. Modern knowledge systems should not replace indigenous systems but should strengthen it. Farmers should be encouraged to use information from various sources to develop solutions that are suitable to their contexts. Government, private institutions and NGOs that aim to improve the food security level of smallholder farmers should build on farmers' local understanding of their environments. Government extension services, which are currently geared towards mainly sharing information that promote adopt modern systems, should be refocused towards promoting both modern and indigenous knowledge systems. This requires that the extension agents be challenged to change their mindset and acknowledge the importance of integrating indigenous knowledge into their training sessions so that what is relevant to farmers may be adopted, enhancing and negotiating the integration process (bettering). Farmers must be capacitated through education and support services to improve and integrate modern practices into their indigenous knowledge systems into current.

While integration of modern and indigenous knowledge is important, it should not be haphazard. It should be such that there will be co-innovation by farmers with their indigenous knowledge and scientists with their scientific knowledge. It should go from the known to the unknown. That will be easier and better achieved using the "plug-in" process (Dittoh 2025 Chap. 1) especially as indigenous processes are largely systems-oriented while available agriculture scientific knowledge is largely commodity-oriented. It is easier to plug commodity-oriented production into a systems-oriented one than vice-versa. This research shows that modern agricultural knowledge emphasizes on increase in yields and mono-cropping. The objectives of agriculture production, however, goes beyond yields to improvements in human nutrition, soil health, animal health and other considerations, which are vital for the achievements of SDG2. The evidence in this research shows that integration of indigenous and modern systems is the way to go; but the results also indicate and imply that, if costs, availability of inputs, nutrition and environmental and other factors are considered, it will be better to start the process of integration by plugging scientific knowledge into available indigenous knowledge.

References

Adekunle AA, Fatunbi AO (2012) Approaches for setting-up multi-stakeholder platforms for agricultural research and developmente. World Appl Sci J 16(7):981–988

Adesina AA, Mbila D, Nkamleu GB, Endamana D (2000) Econometric analysis of the determinants of adoption of alley farming by farmers in the forest zone of southwest Cameroon. Agriculture, ecosystems & environment 80(3):255–265

Agrawal A (1995) Dismantling the divide between indigenous and scientific knowledge. Dev Chang 26(4):13–39

Agrawal A (2009) Why "indigenous" knowledge? J R Soc N Z 39(4):157–158. https://doi.org/10.1080/03014220909510569

Ainembabazi JH, Mugisha J (2014) The role of farming experience on the adoption of agricultural technologies: evidence from smallholder farmers in Uganda. J Dev Stud 50(5):666–679. https://doi.org/10.1080/00220388.2013.874556

Ajewole OO, Ayinde OE, Ojehomon VET, Agboh-Noameshie RA, Diagne AA (2015) Gender analysis of agricultural innovation and decision making among rice farming household in Nigeria. J Agric Inform 6(2):72–82. https://doi.org/10.17700/jai.2015.6.2.179

Amsalu A, de Graaff J (2007) Determinants of adoption and continued use of stone terraces for soil and water conservation in an Ethiopian highland watershed. Ecol Econ 61(2–3):294–302. https://doi.org/10.1016/j.ecolecon.2006.01.014

Apraku A, Morton JF, Apraku Gyampoh B (2022) Climate change and small-scale agriculture in Africa: does indigenous knowledge matter? Insights from Kenya and South Africa. Sci Afr 12:e00821. https://doi.org/10.1016/j.sciaf.2021.e00821

Armbruster H, Bikfalvi A, Kinkel S, Lay G (2008) Organizational innovation: the challenge of measuring non-technical innovation in large-scale surveys. Technovation 28(10):644–657. https://doi.org/10.1016/j.technovation.2008.03.003

Arslan A, McCarthy N, Lipper L, Asfaw S, Cattaneo A (2014) Adoption and intensity of adoption of conservation farming practices in Zambia. Agric Ecosyst Environ 187:72–86. https://doi.org/10.1016/j.agee.2013.08.017

Asfaw S, Shiferaw B, Simtowe F, Lipper L (2012) Impact of modern agricultural technologies on smallholder welfare: evidence from Tanzania and Ethiopia. Food Policy 37(3):283–295. https://doi.org/10.1016/j.foodpol.2012.02.013

Basdew M, Jiri O, Mafongoya PL (2017) Integration of indigenous and scientific knowledge in climate adaptation in KwaZulu-Natal, South Africa. Change Adapt Socio-Ecol Syst 3(1):56–67. https://doi.org/10.1515/cass-2017-0006

Bolosha A, Sinyolo S, Ramoroka KH (2023) Factors influencing innovation among small, micro and medium enterprises (SMMEs) in marginalized settings: Evidence from South Africa. Innovation and Development 13(3): 583–601

Biosafety, A. African C. for (2012) South Africa's seed systems: challenges for food sovereignty. African Centre for Biosafety (ACB)

Briggs J (2005) The use of indigenous knowledge in development: problems and challenges. Prog Dev Stud 5(2):99–114. https://doi.org/10.1191/1464993405ps105oa

Briggs J, Moyo B (2012) The resilience of indigenous knowledge in small-scale African agriculture: key drivers. Scott Geogr J 128:64–80

Buthelezi NN (2018) Application of soil indigenous knowledge in rural communities of eastern South Africa. Unpublished PhD thesis, University of KwaZulu-Natal

Buthelezi NN, Hughes JC (2014) Indigenous knowledge systems and agricultural rural development in South Africa: past and present perspectives. Indilinga-Afr J Indigenous Knowl Syst 13(2):231–250

Buthelezi NN, Hughes JC, Modi AT (2013) The use of scientific and indigenous knowledge in agricultural land evaluation and soil fertility studies of two villages in KwaZulu-Natal, South Africa. Afr J Agric Res 8(6):507–518. https://doi.org/10.5897/AJAR11.1108

Cele L, Wale E (2018) The role of land-and water-use rights in smallholders' productive use of irrigation water in KwaZulu-Natal, South Africa. Afr J Agric Resour Econ 13(311–2019–686):345–356. file:///F:/Spec 2/Traffic Delay Model.pdf

Chowa C, Garforth C, Cardey S (2013) Farmer experience of pluralistic agricultural extension, Malawi. J Agric Educ Ext 19(2):147–166. https://doi.org/10.1126/science.38.975.331

Clemens I (2021) The emergence of innovations through the encounter of knowledges in "the local". How fresh action emerges in networks. Int J Train Dev 25(4):402–413. https://doi.org/10.1111/ijtd.12241

Cousins B (2013) Smallholder irrigation schemes, agrarian reform and "Accumulation from Above and from Below" in South Africa. J Agrar Chang 13(1):116–139. https://doi.org/10.1111/joac.12000

Conley TG, Udry CR (2010) Learning about a new technology: Pineapple in Ghana. American Economic Review 100(1):35–69

Cozzens S, Sutz J (2014) Innovation in informal settings: reflections and proposals for a research agenda.

Innov Dev 4(1):5–31. https://doi.org/10.1080/21579
30x.2013.876803

Davis M, Laas K (2014) "Broader Impacts" or "Responsible Research and Innovation"? A Comparison of Two Criteria for Funding Research in Science and Engineering. Sci Eng Ethics 20(4):963–983. https://doi.org/10.1007/s11948-013-9480-1

de Bont C, Komakech HC, Veldwisch GJ (2019) Neither modern nor traditional: Farmer-led irrigation development in Kilimanjaro Region, Tanzania. World Development 116:15–27

Diiro GM, Seymour G, Kassie M, Muricho G, Muriithi BW (2018) Women's empowerment in agriculture and agricultural productivity: evidence from rural maize farmer households in western Kenya. PloS One 13(5):1–27. https://doi.org/10.1371/journal.pone.0197995

Dirwai TL, Senzanje A, Mudhara M (2019) Water governance impacts on water adequacy in smallholder irrigation schemes in KwaZulu-Natal province, South Africa. Water Policy 21(1):127–146

Dittoh S (2003) Improving availability of nutritionally adequate and affordable food supplies at community levels in West Africa. In: 2nd international workshop: food-based approaches for a healthy nutrition, pp 23–28

El-Hani CH, Poliseli L, Ludwig D (2022) Beyond the divide between indigenous and academic knowledge: causal and mechanistic explanations in a Brazilian fishing community. Stud Hist Philos Sci 91:296–306

Ellis F, Kutengule M, Nyasulu A (2003) Livelihoods and rural poverty reduction in Malawi. World Dev 31(9):1495–1510. https://doi.org/10.1016/S0305-750X(03)00111-6

Gatzweiler FW, Von Braun J (2016) Technological and institutional innovations for marginalized smallholders in agricultural development. Springer Nature

Gebre GG, Zegeye DM (2014) Challenges of farmers' innovativeness in central zone, Tigray, Ethiopia. Int J Agric Policy Res 2(May):215–223

Ghimire R, Huang WC (2015) Household wealth and adoption of improved maize varieties in Nepal: a double-hurdle approach. Food Secur 7(6):1321–1335. https://doi.org/10.1007/s12571-015-0518-x

Ghimire R, Huang WC (2016) Adoption pattern and welfare impact of agricultural technology: empirical evidence from rice farmers in Nepal. J South Asian Dev 11(1):113–137. https://doi.org/10.1177/0973174116629254

Gizaw B (2003) Blending of traditional and modern technologies through science. In: International Conference on African Development Archives. http://scholarworks.wmich.edu/africancenter_icad_archivehttp://scholarworks.wmich.edu/africancenter_icad_archive/67%0Ahttp://scholarworks.wmich.edu/africancenter_icad_archive%0Ahttp://scholarworks.wmich.edu/africancenter_icad_archive/67

Godin B (2006) The linear model of innovation: the historical construction of an analytical framework. Sci

Technol Human Values 31(6):639–667. https://doi.org/10.1177/0162243906291865

Gómez-Baggethun E (2022) Is there a future for indigenous and local knowledge? J Peasant Stud 49(6):1139–1157. https://doi.org/10.1080/03066150.2021.1926994

Gupta AK (2016) Agricultural extension in South Asia grassroots innovation: minds on the December, pp 1–4. https://www.aesanetwork.org/wp-content/uploads/2018/02/GRASSROOTS-INNOVATION-MINDS-ON-THE-MARGIN-ARE-NOT-MARGINAL-MINDS.pdf

Hart T, Mouton J (2005) Indigenous knowledge and its relevance for agriculture: a case study in Uganda. Indilinga African Journal of Indigenous Knowledge Systems 4(1):249–263

Heeks R, Foster C, Nugroho Y (2014) New models of inclusive innovation for development. Innov Dev 4(2):175–185. https://doi.org/10.1080/21579
30x.2014.928982

Hendriks S (2013) Food security in south africa: status quo and policy imperatives aeasa presidential Address 1 October 2013, Bela Bela. Agrekon 52(2):1–24

Hermans TDG, Šakić Trogrlić R, van den Homberg MJC, Bailon H, Sarku R, Mosurska A (2022) Exploring the integration of local and scientific knowledge in early warning systems for disaster risk reduction: a review. Nat Hazards 114(2):1125–1152. https://doi.org/10.1007/s11069-022-05468-8

Von Hippel E (2005) Democratizing innovation: The evolving phenomenon of user innovation. Journal für Betriebswirtschaft 55:63–78

Hiwasaki L, Luna E, Syamsidik, & Shaw, R. (2014) Process for integrating local and indigenous knowledge with science for hydro-meteorological disaster risk reduction and climate change adaptation in coastal and small island communities. Int J Disaster Risk Reduct 10:15–27. https://doi.org/10.1016/j.ijdrr.2014.07.007

Kassie M, Shiferaw B, Muricho G (2011) Agricultural technology, crop income, and poverty alleviation in Uganda. World Dev 39(10):1784–1795. https://doi.org/10.1016/j.worlddev.2011.04.023

Kebebe E (2019) Bridging technology adoption gaps in livestock sector in Ethiopia: a innovation system perspective. Technol Soc 57:30–37. https://doi.org/10.1016/j.techsoc.2018.12.002

Lundvall B-A (2009) Innovation as an interactive process: user-producer interaction to the national system of innovation. Afr J Sci Technol Innov Dev 1(2_3):10–34

Malek MA, Gatzweiler FW, Von Braun J (2017) Identifying technology innovations for marginalized smallholders-A conceptual approach. Technol Soc 49:48–56. https://doi.org/10.1016/j.techsoc.2017.03.002

Malone A, Santi P, Cabana YC et al (2022) Cross-validation as a step toward the integration of local and scientific knowledge of geologic hazards in rural Peru. Int J Disaster Risk Reduct 67(10):26–82

Mariano MJ, Villano R, Fleming E (2012) Factors influencing farmers' adoption of modern rice technologies and good management practices in the Philippines. Agr Syst 110:41–53

Mignouna DB, Manyong VM, Rusike J, Mutabazi KDS, Senkondo EM (2011) Determinants of adopting imazapyr-resistant maize technologies and its impact on household income in Western Kenya. AgBioforum 14(3):158–163

Morris ML (2007) Fertilizer use in African agriculture: lessons learned and good practice guidelines. World Bank Publications

Muchara B, Ortmann G, Wale E, Mudhara M (2014) Collective action and participation in irrigation water management: a case study of Mooi River Irrigation Scheme in KwaZulu-Natal Province, South Africa. Water SA 40(4):699. https://doi.org/10.4314/wsa.v40i4.15

Muchara B, Ortmann G, Mudhara M, Wale E (2016) Irrigation water value for potato farmers in the Mooi River Irrigation Scheme of KwaZulu-Natal, South Africa: a residual value approach. Agric Water Manag 164:243–252

Ncube S, Madikizela LM, Chimuka L, Nindi MM (2018) Environmental fate and ecotoxicological effects of antiretrovirals: a current global status and future perspectives. Water Res 145:231–247

Ndabeni L (2019) Innovation and the dynamics of rural economic development. In: Jacobs PT (ed) Equitable rural socio-economic transitions. HSRC Press, Cape Town, pp 219–229

Ninh LK (2020) Economic role of education in agriculture: evidence from rural Vietnam. J Econ Dev 23(1):47–58. https://doi.org/10.1108/JED-05-2020-0052

Nyong A, Adesina F, Osman Elasha B (2007) The value of indigenous knowledge in climate change mitigation and adaptation strategies in the African Sahel. Mitig Adapt Strat Glob Chang 12(5):787–797. https://doi.org/10.1007/s11027-007-9099-0

Oduro Ofori E, Braimah I, Osei K (2015) Promoting green infrastructure in Kumasi: challenges and prospects promoting green infrastructure in Kumasi: challenges and strategies. Iiste 4:110–119

Okpachu AS, Okpachu OG, Obijesi IK (2014) The Impact of education on agricultural productivity of small scale rural female maize farmers in Potiskum Local Government, Yobe State: a Panacea for Rural Economic Development in Nigeria. Int J Res Agric Food Sci 2(4):26–33

Pamuk H, Bulte E, Adekunle AA (2014) Do decentralized innovation systems promote agricultural technology adoption? Experimental evidence from Africa. Food Policy 44:227–236. https://doi.org/10.1016/j.foodpol.2013.09.015

Phakathi S, Wale E (2018) Explaining variation in the economic value of irrigation water using psychological capital: a case study from ndumo b and Makhathini, KwaZulu-Natal, South Africa. Water SA 44(4):664–673. https://doi.org/10.4314/wsa.v44i4.15

Phakathi S, Sinyolo S, Marire J, Fraser G (2021) Heterogeneous welfare effects of farmer groups in smallholder irrigation schemes in South Africa. Afr J Agric Resour Econ 16(1):27–45. file:///F:/Spec 2/Traffic Delay Model.pdf

Raut N, Sitaula BK, Vatn A, Paudel GS (2011) Determinants of adoption and extent of agricultural intensification in the central mid-hills of Nepal. J Sustain Dev 4(4):47. https://doi.org/10.5539/jsd.v4n4p47

Ray S (2023) Weaving the links: traditional knowledge into modern science. Futures 145:103081. https://doi.org/10.1016/j.futures.2022.103081

Retnowati A, Anantasari E, Marfai MA, Dittmann A (2014) Environmental ethics in local knowledge responding to climate change: an understanding of seasonal traditional calendar PranotoMongso and its Phenology in Karst Area of GunungKidul, Yogyakarta, Indonesia. Procedia Environ Sci 20:785–794. https://doi.org/10.1016/j.proenv.2014.03.095

van Rijn F, Bulte E, Adekunle A (2012) Social capital and agricultural innovation in Sub-Saharan Africa. Agr Syst 108:112–122. https://doi.org/10.1016/j.agsy.2011.12.003

SA S (2019) Statistics South Africa National poverty lines, National Poverty Lines. 12(2):406–421. http://Www.Statssa.Gov.Za/Publications/P03101/P031012019.Pdf. Accessed 23/11/2019

Santiago F (2014) Innovation for inclusive development. Innov Dev 4(1):1–4. https://doi.org/10.1080/2157930X.2014.890353

Scoones I (2009) Livelihoods perspectives and rural development. J Peasant Stud 36(1):171–196. https://doi.org/10.1080/03066150902820503

Selejio O, Lasway JA (2019) Economic analysis of the adoption of inorganic fertilisers and improved maize seeds in Tanzania. Afr J Agric Res Economics 14(4):310–330

Shiferaw B, Bank W, Muricho G, Shiferaw BA (2011a) Farmer organizations and collective action institutions for improving market access and technology adoption in Sub-Saharan Africa: review of experiences and implications for polic. In: Experiments for development view project sustainable land, water and Agr, pp 1–41. https://www.researchgate.net/publication/281321739

Shiferaw B, Hellin J, Muricho G (2011b) Improving market access and agricultural productivity growth in Africa: what role for producer organizations and collective action institutions? Food Secur 3:475–489

Sinyolo S (2020) Technology adoption and household food security among rural households in South Africa: the role of improved maize varieties. Technol Soc 60:101214. https://doi.org/10.1016/j.techsoc.2019.101214

Sinyolo S, Mudhara M (2018) The impact of social capital on entrepreneurship among smallholder farmers in rural South Africa. J Rural Dev 37:519–538

Spielman DJ, Ekboir J, Davis K (2009) The art and science of innovation systems inquiry: applications to Sub-Saharan African agriculture. Technol Soc 31(4):399–405. https://doi.org/10.1016/j.techsoc.2009.10.004

Suvedi M, Ghimire R, Kaplowitz M (2017) Farmers' participation in extension programs and technology adoption in rural Nepal: a logistic regression analysis. J Agric Educ Ext 23(4):351–371

Technology and Innovation (DSI) (2019) White paper on science, technology and innovation. DSI. Government Gazette, Pretoria, p 41909

Triomphe B, Floquet A, Kamau G, Letty B, Vodouhe SD, Ng'ang'a T, Stevens J, van den Berg J, Selemna N, Bridier B, Crane T, Almekinders C, Waters-Bayer A, Hocdé H (2013) What Does an Inventory of Recent Innovation Experiences Tell Us About Agricultural Innovation in Africa? J Agric Educ Ext 19(3):311–324. https://doi.org/10.1080/1389224X.2013.782181

Ubisi NR, Kolanisi U, Jiri O (2019) Comparative review of indigenous knowledge systems and modern climate science. Ubuntu: J Conflict Soc Trans 8(2):53–73

Wang J (2015) Integrating indigenous with scientific knowledge for the development of sustainable agriculture: studies in Shaanxi Province. Asian J Agric Dev 15:41–58. file:///F:/Spec 2/Traffic Delay Model.pdf

Weir S, Knight J (2014) Externality effects of education: dynamics of the adoption and diffusion of an innovation in rural Ethiopia. Econ Dev Cult Chang 53(1):93–113

Yanou MP, Ros-Tonen M, Reed J, Moombe K, Sunderland T (2023) Efforts to integrate local and scientific knowledge: the need for decolonising knowledge for conservation and natural resource management. Heliyon 9:e21785

Decolonizing Food Systems Through the Plug-In Principle: The Case of Cereal Seed Value Chains in Mali

Anna Bon, Mohamed Coulibaly, André Baart,
Lucas de Lange, and Wendelien Tuijp

Abstract

Food systems transformation is the new idea for food security policy for sub-Sahara Africa, in which *inclusion*, *demand-driven innovation* and *participation* are mentioned as key words. However, inclusion of (indigenous) knowledge and farmer innovation are absent in most food security strategies. Instead, technology transfer and commercialization are promoted as key to food production and poverty reduction. In this Chapter, based on field work in rural West Africa, we explore the case of two food systems in Mali, one (i) based on transfer of technology to increase crop production; the other (ii) based on indigenous knowledge and local practices. It shows how two approaches can co-exist and be combined and merged with existing practices, without dismissing or substituting what is already in place. The study also reveals the advantages of human-centered, context-aware approaches, in which the complex interplay between local practices and contextual factors are considered, and local food sovereignty is achieved.

Keywords

Food security · Food sovereignty · Transfer of technology · Indigenous knowledge · Decision-making

4.1 Food Security Versus Food Sovereignty

In the aftermath of the Covid-19 pandemic and in the light of climate change and armed conflicts, food insecurity is rising in Africa. According to a recent study by the FAO, 346.4 million Africans are suffering from severe food insecurity, while 452 million suffer from moderate food insecurity (FAO 2021). In line with the United Nations' sustainable development goal SDG2 "Zero hunger", numerous food security intervention programmes are being designed, aimed at food systems transformation[1] in sub-Sahara Africa (Sasson 2012, ICRISAT 2021-2025, Kemoe et al. 2022). The focus is on intensifying food production, while reducing carbon emissions (Gautam et al. 2022).

A. Bon (✉) · A. Baart · L. de Lange · W. Tuijp
Vrije Universiteit Amsterdam,
Amsterdam, The Netherlands
e-mail: a.bon@vu.nl

M. Coulibaly
Vrije Universiteit Amsterdam,
Amsterdam, The Netherlands

Université des Sciences Juridiques et Politiques de Bamako, Bamako, Mali

[1] https://www.un.org/en/food-systems-summit/news/making-food-systems-work-people-planet-and-prosperity

S. Dittoh et al. (eds.), *Integrating Indigenous and Scientific Knowledge for Sustainable Food Systems in Africa*, Sustainable Development Goals Series,
https://doi.org/10.1007/978-3-031-85512-2_4

The international policy to ensure *food security*, supported by the World Bank, the United Nations (cf. UN Food System Summit 2021) and many other international development agencies is based on *technology and knowledge transfer*.

Apart from the broad consensus on *food security* policy in the international development community, other alternative approaches exist, that tackle the food challenges from the perspective of local *food sovereignty*. There is a wide discord between the advocates of food sovereignty and the proponents of conventional food security policies (Gerretsen 2022). Despite the similar names, *food security* and *food sovereignty* represent different values and beliefs (Morvaridi 2012). They have different roots and practices and can therefore be considered as two different paradigms. The differences between the two paradigms have extensively been described by many authors in different studies (e.g. Jarosz 2014; Pachón-Ariza 2013; Wald and Hill 2016). In this Chapter we use these two paradigms as a reference to assess and discuss the case of seed value chains in rural Mali. We take, as a practical application reference, the plug-in principle proposed in Chap. 1 of the present book.

Food sovereignty is defined in the literature as the right of peoples to autonomously define their own food and agriculture systems (Declaration of Nyléni 2007, Pachón-Ariza 2013). It includes not only the fundamental human right to healthy food, but also the right to have culturally appropriate food, which is ecologically produced according to sustainable methods (Kiptot et al. 2014). Food sovereignty unifies four pillars, namely (i) the right to food and food sovereignty, (ii) mainstreaming agro-ecological family farming, (iii) defending people's access to and control of natural resources, and (iv) autonomy and trade.

Food and nutrition security is the term used in international development policy. In this paradigm, food security is related to economic growth (Gassner et al. 2019). Complete food security is only achieved when all people have physical, social and economic access to sufficient nutritious food at all times. Lack of inputs (notably improved/hybrid seeds, fertilizers, pesticide), and a lack of access to international markets and small-scale production are considered the root

causes of food shortage. To address these supposed deficiencies, transfer of technologies, capacity building and market development are the proposed actions to improve food security (Canfield et al. 2022; Ali Mohamed et al. 2021).

A point of critique at the seemingly neutral term "food security" (e.g. Ferguson et al. 2022; McKeon 2021), is that the discussion about autonomy, domination and social control of available food systems is silenced. The food security approach fails to answer important questions like (i) where should food be produced, (ii) how, (iii) by whom and (iv) under which conditions, (vi) who is benefiting and (vii) who is in control? (Montenegro de Wit et al 2021).

Since many food sovereignty advocates are critical and aim at changing existing, inequitable, social, political and economic structures and politics that they see as the root cause of social and environmental destruction in both North and South (Wittman et al. 2010), calls for food sovereignty are seen as *decolonial* movements, that should not be included or mentioned in the mainstream development discourse. However, the importance of human-centered, holistic approaches to food systems, in which the complex interplay between local practices and contextual factors is addressed, is stressed in various studies (McKeon 2021; Sanga et al. 2021).

4.2 Structure of This Chapter

In this chapter we explore the case of seed value chains in rural Mali, looking at two different food systems, one based on improved, certified and commercialized cereal seeds, and the other based on so-called *peasant seeds*. Based on our field visits in Mali, interviews and focus groups with local farmers and farmer organizations and using the framework of *food security vs sovereignty* as a reference and the plug-in principle described by Dittoh et al. (2016) as a practical framework, we explore the sustainability of the food systems. In the next section we describe the research methods used in this research project, and that is followed by descriptions of our case study on cereal seed systems in Mali. We describe a conventional approach, supported by various research institutions. We also present an alternative approach, supported by local farmer organizations and based on local practice. In Sects. 4.4 and 4.5 we discuss the case study based on our theoretical framework and in Sect. 4.6 we give conclusions and some recommendations.

4.3 The Research Approach and Methodology

This research is part of an ongoing action research program, named Web Alliance for Regreening in Africa[2] (W4RA) which strives to support innovation by small-holder farmers in the Sahel through co-design and deployment of socio-technical solutions to improve communication and access to information. Through this research program, we have been exposed to cases related to food security/food sovereignty in the Sahel. The research methods we used were field visits, collaborative design workshops with seed producers, focus group discussions with farmers and other stakeholders in the seed value chain, interviews with local agriculture experts and various group discussions, presentations and conceptual modeling and co-design with the stakeholders during our stay in Mali. Our design-science tools to analyze patterns is to map stakeholders and their goals using stakeholder analysis or strategy-goal conceptual modeling techniques (see Figs. 4.1 and 4.2).

4.3.1 A Case Study of Seed Value Chains in Rural Mali

In our case study concerning the seed value chain in Mali, we worked between 2015 and 2024, together with the *Association des Organisations Professionnelles Paysannes* (AOPP), the national umbrella association for professional farmer organizations in Mali. The work presented in this chapter is the result of four weeks of workshops with AOPP including four field visits, in April 2019, January 2020, October 2021, February

2022 and November 2023. We interviewed seed-producing smallholder farmers and visited them in their fields. We visited seed cooperatives and seed trading unions. We also visited the seed certifying national laboratory, Labosem. We interviewed commercial seed traders. We have been assisted in our research by local radio journalists and various Malian and Burkinabe experts in rural development and local entrepreneurship.

AOPP's activities cover the whole of Mali. AOPP reaches up to three million people in rural regions — and is concerned with the trading position of its member cooperatives or OPs (*organisations paysannes*). AOPP supports directly up to 250 local farmer organizations and represents them at the national level trying to enhance communication and provide them with timely, accurate and relevant information from all segments of the seed value chain. The case study in Mali shows the contrast between approaches towards food security, and food sovereignty.

4.3.2 Comparing Two Seed Ecosystems in Mali

Mali is situated in West Africa and has a population (in 2020) of about 19.5 million people of whom 75% live on agriculture. Pearl millet (Pennisetum glaucum) and sorghum (Sorghum bicolor) are the two main cereal crops cultivated in the arid and semi-arid areas of Mali (Coulibaly et al. 2014). The ecosystem for cereal seeds in Mali is a complex one. On one hand there is a traditional farmer-managed seed system in which smallholders exchange cereal seeds often even without monetary transactions: this is called the peasant seed system.[3] The core of this system is small scale (subsistence) farming and bartering, using natural seeds, but it does not lead to much surpluses or profit. There is another, commercial system of improved seeds that was first introduced in the late 1990s and early 2000s in

---

[2]See also https://w4ra.org Accessed 10 June 2024.

[3]The information about the traditional seed system is obtained by a long interview with Barke Ousmane Diallo from AOPP and a collaborative workshop at AOPP's office in Bamako, January 2021),

Fig. 4.1 Challenges of the commercial seed system identified during group discussions, AOPP office, Bamako, January 2020

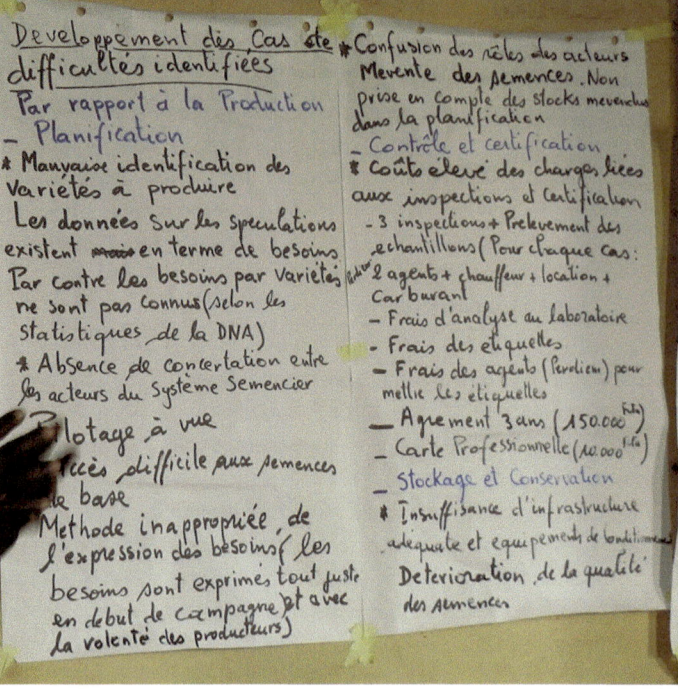

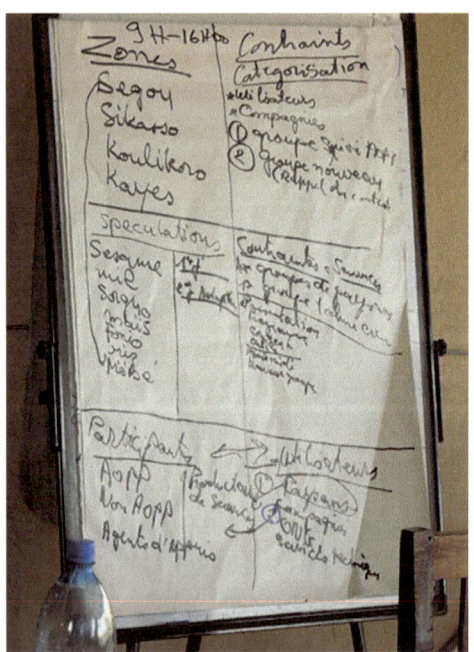

Fig. 4.2 Summary of identified barriers to the adoption of certified seeds, January 2022

Mali by international NGOs and international research institutes with support by the Malian government.

The traditional or *peasant seed system*, usually result in considerably lower yields than the laboratory (hybrid) seeds. Still, many farmers rely on them, as they are available without investments, and no pesticides are necessary. They are more resistant to drought and can be, in contrast to the hybrid seeds that only breed two or three generations, replanted/reused without limitations. Peasant seeds are at the basis of daily food consumption for large part of the population in Mali and form the backbone of Malian agriculture, making up about 95% percent of the local food consumption. The commercial system, in contrast to the traditional system which has as the main purpose to feed the family or the community, is set up by the Malian government and international NGOs to boost agriculture production through modernization and trade by introducing hybrid, laboratory seeds that have much higher yields. The production of the high yielding seeds is supposed to give farmers better income and ensure that Mali adequately copes with a rapidly growing population and produces staple food to feed the country. Moreover, the production surplus is supposed to lead to a more effective national food ecosystem, that gives traders new opportunities to export seeds to neighboring countries. This is supposed to lead to a more efficient and effective agricultural sector and consequently to economic growth, which in turn will trickle down to the beneficiaries at the bottom of the pyramid.

4.4 Summary of the Food Systems Analysis Workshops in Bamako

During a co-design workshop in January 2020 in Bamako, we did a thorough analysis of the advantages and barriers of the seed value chains in Mali. We invited 40 seed-producing farmers to the workshop at AOPP. The method of analysis was collaborative exploration of the problem space and subsequently a solution space. The workshop was also a good opportunity for people to meet and discuss in an open and creative way. During the workshop we discussed the preoccupations of the seed producers, the political context in Mali, and the need for a better information & communication system to improve the seed networks. The regions of intervention for the seeds by AOPP are: Ségou, Sikasso, Koulikoro and Kayes. The crops concerned are sorghum, millet, maize, rice, niébé (beans), sesame and fonio.

4.4.1 Advantages of Indigenous Knowledge-Based Peasant Seeds

AOPP is concerned with the future of the peasant seeds in Mali. About 80% of all food is produced from traditional seeds, based on indigenous farming knowledge. Only about 20% is from (improved, hybrid) certified seeds. The traditional seeds are commonly exchanged in local markets, without reaching the official commercial value chains. The following are advantages of the traditional, unimproved agro-ecological seeds:

- No investment costs for farmers as there is no need for fertilizers, pesticides and certification. The peasant seeds do not require these.

- Better for biodiversity as no pesticide is used, and no insects that are important for pollination and to feed birds (local poultry) are killed; biodiversity in general is preserved.
- Better for the soil as the seeds (of multiple crops) are well adapted to the soil, and do not exhaust the soil as the monocultures of improved seeds.
- Traditional seeds are adaptive and can better cope with changing climatic conditions such as erratic rainfall patterns, extreme temperatures etc.
- Crowd-sourcing system for joint/shared investments can be implemented by the farmers (SPG: système participative de garantie).
- Minimum complexities compared to certified seeds requirements; e.g. risk of depreciation due to contamination of crops with other seeds, lack of microfinances to pre-finance, prize fluctuations leading to`mévente´ unsold produce or against a lower price than expected.

Despite the advantages of indigenous knowledge-based peasant seeds, the hybrid, improved seeds can be advantageous to some farmers and to Mali, with respect to increases in productivity (yields). There are, however, challenges with certified seeds. They include: (i) they are not profitable for farmers if all production costs are taken into account. (ii) the certified seed market is a middle-man dominated market in which the (poor) farmers carry all risks and have limited access to market information (iii) certification is a time-consuming, beaucratic process that hampers the value exchange process.

During focus group discussions with seed producing farmers in the Segou, Sikasso, Kolikoro and Kayes regions of Mali, local farmer organizations reported to have adopted the improved cereal seeds because they are useful to farmers from a commercial point of view. The improved seeds give higher crop yields and can be sold in official commercial and global markets. They give farmers the opportunity to invest and earn financial revenues. However, the farmers also admit that, until present day, the commercial seed system is inefficiently organized. Seed producers—usually smallholders—exchange, buy, produce and sell seeds through seed cooperatives and unions, but the supply and demand do not match. The system is hampered by lack of good channels for communication between producers and buyers to enable informed decision-making by farmers of what to cultivate. Poor farmers in regions of rain-fed agriculture are especially vulnerable to price and demand fluctuations. Figure 4.1 above shows the challenges of the commercial seed system identified during the group discussion.

Before the improved (hybrid) seeds are marketed they must undergo official certifying process. This certification process is led by a national agency. Ìt is a prolonged and bureaucratic process. Here is an account of our visit to the national seed certifying lab:

Labosem – Laboratoire des Semences in Bamako is the institute where seeds are tested and certified. Our research includes a visit and interview with the director of Labosem, *Chef de Laboratoire, national expert on improved seed quality* on 8 october 2021. *LABOSEM is part of the Direction Nationale de l'Agriculture (DNA) of the Malian Ministry of Agriculture. It is responsible for training and implementation of the Law of 2010, named:*

Loi 10–032 du 12 juillet 2010 relative aux semences d'origine végétale. "La présente loi fixe les règles de gestion, de production, de commercialisation et de contrôle de qualité des semences d'origine végétale (les semences issues de variétés améliorées ou traditionnelles). Les variétés traditionnelles constituent un patrimoine national. Elles doivent être gérées dans l'intérêt de la nation et conformément aux conventions internationales ratifiées par le Mali. Cependant, les variétés créées sont la propriété des obtenteurs." The responsibilities of LABOSEM are: Certification, Inspection of Genetic resources, Authorization/permit, Registration, Plant production, Plant protection, Plant variety, Planting material/seeds, Processing/handling, Packaging/labelling. IER: Institut de l'Economie Rurale produces and provides the seeds "base" and "pre-base".

For export of seeds between neighboring ECOWAS countries, seeds need to be certified, this is set in a harmonized ECOWAS law. Currently in Mali, only 5% of all crop seeds are certified seeds. They are in generally bought by international development aid organizations: FAO, Red Cross, USC Canada etc. There is hardly a local market for certified seeds. Peasant seeds constitute the other 95% used for food production in Mali. They are locally traded as non-monetized exchange produce. The market for certified seeds in Mali suffers from "mévente", i.e. the offer (supply) largely exceeds the market demand. There is therefore an imbalance and the market is ineffective and inefficient. While the price per kilogram of certified seeds is 300 fCFA, it is often sold as food for half of the price per volume. After being certified an official paper is given. The process of certification of seeds can take 1–3 months. It is only with the certification documents that the improved seed producing farmers can start sales. Sometimes the certificates get lost during transport. AOPP is now helping to transfer the certificates to avoid loss and/or delays.

During our meeting, the director of Labosem explained that farmers need to clear their fields as to obtain certification of their produced seeds. We asked if that also counts for trees. In the policy, it says yes. The discussion went into the direction of agro-ecology, regreening and FMNR, which point to the opposite direction (agro-forestry, mixed crops/intercropping, protection of biodiversity, biological farming etc.). This is obviously not the area of interest to those promoting certified seed production.

From the discussion at Labosem and the workshops with seed producers in both commercial and traditional seeds at AOPP, it became clear that some opposing views exist between the proponents of the commercial, certified seed systems, which include the Malian government, international NGOs and commercial seed traders on one side and the proponents of sustainable agriculture, biodiversity conservation and the importance of trees and natural vegetation on farmlands in Africa on the other side. In terms of environmental conservation, peasant seeds are currently gaining attention. However, many food security projects led by research institutes and international organizations have not embraced peasant seeds as a practice that should be incorporated or integrated and diffused as a local innovation.

At AOPP, opinions are divided about the advantages of hybrid versus peasant seeds. Some farmers, members of AOPP, have switched to hybrid improved seeds because they clearly see the improvement in terms of crop yields. Certification of these seeds also opens opportunities to them for trade, access to national and international buyers and even export. Use of certified seeds requires very conscious approach to farming including land use, and application of fertilizers and pesticides. This is not possible for many smallholder farmers: men and women who do not have the organizational structures and financial capacity in place to switch to the use of certified improved/hybrid seeds. Improved seeds come with a list of difficult requirements related to farming and production: clearing of fields, certifying of seeds, cutting of trees, use of pesticides, fertilizers, which implies finance, posing many risks for farmers given the unreliable rains and harvests. Definitively, the two seed systems in Mali do not coexist without friction, leaving many issues open for debate. They are not only different in their processes but also in their core goals. The traditional system is about subsistence, while, the commercial system is aimed at increasing crop yield and boosting commercial seed trade companies. This friction is not caused by the farmers, but seems to be driven by national and international policies, actions, and interests to promote the usage of improved seeds. For example, the international research institute, ICRISAT, mentions their goal as: "helping move farmers from subsistence to commercial operators".

It is difficult, from the above analysis, to say what the future will be of the two parallel seed value chains/systems in Mali. The question that remains is whether the right (national and international development) goal is to urge all farmers to become commercial seed producers. Would this be sustainable in the long run and beneficial for their communities and to the local economy? And is the entire project too much steered by unjustifiable commercial (read western) assumptions? However, independent of whether hybrid/commercial seed production should be implemented by all farmers, this field research showed that farmers in rural Mali are not yet fully ready for large-scale international commercial seed trade. A lack of communication channels, access to information, and (digital) technology and a lack of financial resources form significant bottlenecks. It is, however, important to note that Malian (African) smallholder farmers are not subsistence farmers. They produce to feed themselves alright, but also to feed millions of people residing in cities and towns.

4.5 Discussion

The improved seeds are based on the idea of food security. With a rapidly growing population of 3% per year, the population of Mali may double in 20 years. It, therefore, seems reasonable to introduce technologies and innovate agriculture to increase crop yields and encourage commercialization as rapidly as possible, as being promoted by international development agencies such as the World Bank, the FAO and agricultural research institutes such as ICRISAT and other CGIAR centres.

While various member OPs of the AOPP have successfully embraced the certified seeds system to provide them with extra incomes, many others have experienced losses because of the inefficiency of the system, and the fact that majority of farmers are not ready to abandon the traditional seed systems that provide them with the resilience needed to feed their families in case of price fluctuations, failure in obtaining the certification and several other risks. Some studies, for example Coulibaly et al. (2014), propose as the most efficient alternative system, simultaneously dissemination of hybrid seeds while keeping the conservation of agro-biodiversity through the traditional system. That is, combining elements of the formal and informal systems.

Field research and stakeholder analysis with local farmers and communities make clear that the commercial seed systems for certified seeds do not complement many local practices. The commercial seed systems do not consider the way peasant seeds are produced and traditionally exchanged between farmers and communities, mainly without money. Also, they do not consider the resilience of locally grown seeds, the biodiversity aspects that come with them, the advantages for the soil humidity and fertility of mixed crops, and the advantages of managing the natural vegetation and keeping and exploiting trees on farmlands. Peasant seeds represent a locally driven approach based on tradition and local innovations. They are therefore much more aligned with the goals to achieve food sovereignty, which means that the decision-making about the land and its exploitation should be in the hands of local communities.

4.6 Decolonizing Food Systems for Sustainable Development

When all pros and cons are balanced, a picture emerges that peasant seeds represent an important practice for rural communities that cannot and should not be rejected or even downplayed by international development programs. As Coulibaly et al. (2020) wrote: while seed sovereignty movements around the globe are advancing a discourse of repossession, seed activism in Mali is in its early stages. Up to the present day, farmers have not been criminalized for selling uncertified seeds and using seed protected by the plant variety protection, so the two systems co-exist peacefully. The question is what will the near future bring? Are insights at the international and national levels, in the light of climate change and the call for biodiversity, changing their minds on the formerly undisputed benefits of conventional large-scale agriculture? Or are policymakers, heavily involved with the lobby of the international private food sector and related philanthropic institutions that the local interests and autonomy are exchanged for large-scale interventions to the interest of the influential international private stakeholders? It will require much field-based research, national and international debates and participatory action research, to rethink large-scale food production policies and related interventionist approaches. The current renewed global interest in biodiversity, climate change mitigation and regreening might better support the ideas of indigenous knowledge and peasant seeds and bring the fundamental

right to food sovereignty back on the international development policy agendas. But courage, strong networks of grassroots organizations are needed, to make all voices heard, so as to decolonize the food security paradigm of top-down interventionism and technology transfer as a silver bullet for food systems in Africa.

References

Ali Mohamed EM, Alhaj Abdallah SM, Ahmadi A, Lucero-Prisno DE (2021) Food security and COVID-19 in Africa: implications and recommendations. Am J Trop Med Hyg 104(5):1613–1615. https://doi.org/10.4269/ajtmh.20-1590

Canfield M (2022) The ideology of innovation: philanthropy and racial capitalism in global food governance. J Peasant Stud 50:1–25

Coulibaly H, Bazile D, Sidibé A (2014) Modelling seed system networks in Mali to improve farmers seed supply. Sustain Agric Res 3:18–32

Coulibaly M, Claeys P, Berson A (2020) The right to seeds and legal mobilization for the protection of peasant seed systems in Mali. J Human Rights Pract 12(3):479–450

Canfield M, Anderson MD, McMichael P (2021) UN Food Systems Summit 2021: dismantling democracy and resetting corporate control of food systems. Front Sustain Food Syst 5:661552

Dittoh S, Mudhara M, Weobong CA, Muwaya S, Mahdi M (2016). Contributing to global environmental benefits. In Community Innovations in Sustainable Land Management (pp. 174–181). Routledge

FAO, ECA and AUC (2021) Africa – Regional overview of food security and nutrition 2021: statistics and trends. Accra, FAO

Ferguson CE, Green KM, Swanson SS (2022) Indigenous food sovereignty is constrained by time imperialism. Geoforum 133:20–31

Gassner AD, Harris K, Mausch AT, Lopes C (2019) 5, RF Finlayson 6 and P Dobie 4, 2019. Outlook Agric 48(4):309–315

Gautam, M., Laborde, D., Mamun, A., Martin, W., Piñeiro, V.& Vos, R. (2022) Repurposing agricultural policies and support - options to transform agriculture andfood systems to better serve the health of people, economies, and the planet. Policy Report World Bank and IFPRI, pp. 1–57

Gerretsen I (2022) Africa food crisis: Bill Gates and smallholders see different solutions. Climate Home News. https://www.climatechangenews.com/2022/09/08/africa-food-crisis-bill-gates-and-smallholders-see-different-solutions/

ICRISAT Strategic plan 2021–2025. https://www.icrisat.org/wp-content/uploads/2021/08/Strategic-Plan-2021.pdf

Jarosz L (2014) Comparing food security and food sovereignty discourses. Dialogues Human Geograp 4(2):168–181

Kiptot E, Franzel S, Degrande A (2014) Gender, agroforestry and food security in Africa. Curr Opin Environ Sustain 6:104–109. https://doi.org/10.1016/j.cosust.2013.10.019

Kemoe L, Mitra P, Okou CD, Unsal F (2022) How Africa can escape chronic food insecurity amid climate change. International Monetary Fund IMF Blog. https://www.imf.org/en/Blogs/Articles/2022/09/14/how-africa-can-escape-chronic-food-insecurity-amid-climate-change

McKeon N (2021) Global food governance. Development 64(3–4):172–180

Montenegro de Wit M, Canfield M, Iles A, Anderson M, McKeon N, Guttal S, Gemmill-Herren B, Duncan J, van der Ploeg JD, Prato S (2021) Resetting power in global food governance: the UN food systems summit. Development 64:153–161

Morvaridi B (2012) Capitalist philanthropy and the new green revolution for food security. Int J Soc Agric Food 19(2):243–256

Pachón-Ariza FA (2013) Food sovereignty and rural development: beyond food security. Agronomía Colombiana 31(3):362–377

Sasson A (2012) Food security for Africa: an urgent global challenge. Agric Food Secur 1(1):1–16

Sanga U, Sidibé A, Schmitt Olabisi L (2021) Dynamic pathways of barriers and opportunities for food security and climate adaptation in Southern Mali. World Development 148(2021):105663., ISSN 0305-750X. https://doi.org/10.1016/j.worlddev.2021.105663

Wald N, Hill DP (2016) Rescaling alternative food systems: from food security to food sovereignty. Agric Hum Values 33:203–213

Wittman H, Desmarais A, Wiebe N (2010). The origins and potential of food sovereignty. Food sovereignty: Reconnecting food, nature and community, 2

Men and Women in Farmer-Led Irrigation: The Case of the Upper East Region of Ghana

5

Mercy Apuswin Abarike, Saa Dittoh, and Lesley Hope

Abstract

Food and nutrition insecurity continues to be a threat to the people in the arid and semi-arid parts of Ghana, especially under present climate change realities. Irrigated agriculture in such areas is therefore necessary to supplement rainfed production. Farmer-led irrigation (FLI) has been shown to have great potential in the Upper East Region of Ghana, thus its development is very necessary. For FLI to have the expected impact, men and women must play complementary and mutually rewarding roles. The research that resulted in this paper sought to identify the critical constraints and challenges both genders face in irrigated production and to proffer some solutions. The results show the subordinate role women play in irrigated agriculture in the study area because of several socio-economic challenges. The recommendation is that appropriate agroecological practices in irrigation will free irrigators, especially the women, from the perennial issue of lack of resources to purchase expensive imported inputs. There is also need for significant governmental interventions to provide water lifting devices that are convenient for women.

Keywords

Farmer-led irrigation · Gender dynamics · Sustainable development goals · Agroecological practices · Upper East Region (Ghana)

5.1   Introduction

Achieving food and nutrition security (FNS) in Africa, especially West Africa, has been a difficult struggle over several decades. According to FAO et al. (2022), about 42% of the world's population cannot afford healthy diets and it is about 80% for Africa and over 85% for West Africa. With this grim picture of FNS in Africa, it is very doubtful the United Nations' Sustainable Development Goal (SDG) 2 which aims at ending hunger, achieving food security and improved nutrition, and promoting sustainable agriculture by 2030 (UN 2016) can be achieved. The continental and regional targets contained in the Malabo declaration, Agenda 2063, and others (AU 2013) are also clearly at risk, especially given the prevailing and projected climate change realities and the threat

M. A. Abarike (✉) · S. Dittoh
University for Development Studies, Tamale, Ghana

L. Hope
University of Energy and Natural Resources, Sunyani, Ghana

© The Author(s) 2025
S. Dittoh et al. (eds.), *Integrating Indigenous and Scientific Knowledge for Sustainable Food Systems in Africa*, Sustainable Development Goals Series,
https://doi.org/10.1007/978-3-031-85512-2_5

57

of increased desertification (Dietz et al. 2004; Niang et al. 2014). There is an urgent need to find avenues to accelerate food production in quantity and quality. Irrigated agriculture is a major climate change adaptation measure and has great potential for accelerating food production. Attempts at formal irrigated agriculture since the 1970s in West Africa have been dismal failures (Dittoh 1991; Namara et al. 2011; Ofori et al. 2010. Dittoh et al. 2013, Higginbottom et al. 2021). It is thus reasonable to prioritize the development of farmer-led irrigation on the continent. Dittoh (2020) assessed farmer-led irrigation development (FLID) in Ghana and stated that there are about 224,584 hectares under FLI in the country of which 212,004 hectares (more than 94 percent) are small scale (5 ha or less)[1] and the largest area is in the Upper East Region (about 21.11%). It is instructive that while "modern" irrigation systems in West Africa, constructed with enormous resources, fail to increase food production to any appreciable level, smallholder FLI systems across the region have been performing relatively better with virtually no institutional or donor support (Giordano et al. 2012). According to Ayo (2020), "there is considerable evidence that farmer-controlled small-scale irrigation has a better performance record than government-controlled small-scale irrigation". The poor performance of large-scale irrigation systems and government-controlled small-scale systems is basically the issue of not building on indigenous knowledge and practices (FLI) but trying to replace the farmers' irrigated knowledge and expertise. Plugging-in to indigenous knowledge with available appropriate climate smart irrigation technologies is what farmer-led irrigation development (FLID) should be aiming at.

Women have been very important in agriculture in developing countries generally, but more so in Africa. According to Ogunjimi and Adekalu (2002) "small-scale irrigation (Fadama) plays a key role in the economics of Nigeria as a basic source of food, income, and employment, especially for women in the "slack" period of rainfed agriculture". In Ghana, women play very strategic roles in on-farm activities and in food processing, cottage industries and food marketing. In the Upper East Region of the country, especially in the Bawku West, Garu and Tempane Districts, women are very prominent in the FLI value chain, from production through marketing to consumption. It has, however, been stated that "there are intensifying gender inequalities in agriculture that are getting worse because of climate and international development-oriented commercial agriculture interventions" (Vercillo 2021). This statement points to the need to be more gender conscious in FLID to avoid the current situation where donor interventions in agriculture have had very little impact on women empowerment in agriculture.

Farmer-led irrigation development (FLID) is a process where farmers assume a driving role in improving their water use for agriculture by bringing about changes in knowledge production, technology use, investment patterns and market linkages, and the governance of land and water mostly as individuals, but sometimes in small groups (Lefore et al. 2019; Woodhouse et al. 2017). The process is initiated, managed, and financed by farmers themselves. As stated earlier, farmers have done that quite well and to the best of their abilities over time in West Africa. According to Ofosu (2011), the irrigation technologies in FLI are highly productive, and they achieve high profit margins and provide income opportunities to the wider society in terms of labour. They also have relatively high women participation compared to formal irrigation systems. Farmer-led irrigators cultivate relatively small farm sizes which are better managed because farmers provide adequate water and crop nutrients, resulting in high crop yields. A major challenge of FLI in Ghana is the non-recognition and support by the government, unlike in Nigeria, where the World Bank assisted Fadama project is based on FLI practices of Nigerian smallholder irrigators. Fadama, which is a Hausa word for irrigated areas in low-lying wetlands, are traditionally used for irrigation and fishing as well as provision of feed and water for livestock (Ayo 2020). Nigeria's Fadama projects "plugged-in" into the traditional fadama concept, hence its relative success (Mohan 2002).

The aim of this Chapter is to report on men and women irrigation activities in different FLI types

---

[1]Over 80% of the small-scale irrigated farms are 2 ha or less.

in the Upper East Region of Ghana. The objective of the research on which the article is based was to undertake basic socio-economic analysis of irrigated vegetable production by men and women in the Upper East Region of Ghana with a main purpose of identifying the critical constraints and challenges both genders face in production and to proffer some solutions. Complementary production and decision-making roles of men and women at the household level are critical for increase in food production, especially in irrigated agriculture, and for ensuring desirable food systems at the household, community and other levels. Irrigated agriculture is intense, requiring greater attention and special marketing skills, and those qualities tend to be found more in women than men. Men are however more skilled in digging the wells and making mounds and basins. Women continue to be disadvantaged in several ways and this kind of disaggregated analysis helps to unearth known and "hidden" constraints, and together with both men and women co-create solutions.

5.2 The Study Area and Methodology of Research

The study was conducted in the Upper East Region of Ghana, which is located in the northeastern corner of the country, between longitude 00 and 10 W and latitudes 100 30″ and 110 N. It is bordered to the north by Burkina Faso, the east by the Republic of Togo, the west by the Sisala East Municipal of the Upper West Region and the south by West Mamprusi Municipal of North East Region. The land is relatively flat with a few hills to the east and southeast. It is the region with the largest and most advanced irrigation infrastructure in Ghana (Dittoh et al. 2013). It is characterized by shallow and accessible groundwater resources (Anayah et al. 2013; Obuobi et al. 2013). Figure 5.1 shows a map of the Upper East Region and the study sites.

The study aimed at covering the main FLI types in the region. Dittoh et al. (2013) identified four dominant types in the region based on the source of water and method of water delivery onto the irrigated fields. The four identified irrigation types formed the basis for the selection of participants for the study.

Five districts, Tempane, Bawku West, Talensi, Kassena-Nakanna and Builsa North, were purposively selected based on the dominance of FLI in those districts. Cluster sampling technique was used to select the farmer-led irrigators for interview and focus group discussions (FGDs). A semi-structured interview questionnaire was used to obtain basic socio-economic information of men and women irrigators as well as FLI farming systems (i. e. the crops and crop mixtures cultivated). Field observations were also undertaken.

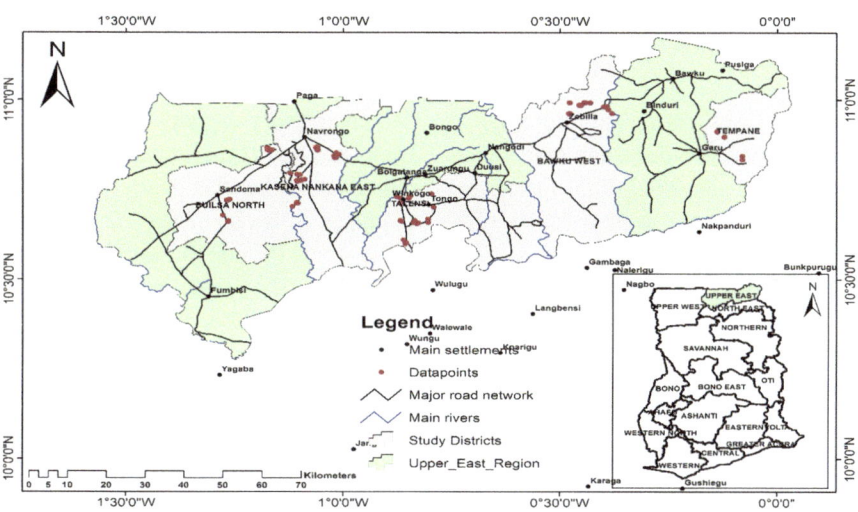

Fig. 5.1 Map of the Upper East Region showing the study areas

Table 5.1 Numbers of respondents according to FLI type

FLI irrigation type[a]	Number of men irrigators	% Freq.	Number of women irrigators	% Freq.	Total	% Freq
(1) Gravity from surface water	19	9.90	8	13.79	27	10.80
(2) Manual (bucket-fetch) from surface water and/or groundwater (wells)	25	13.02	20	34.48	45	18.00
(3) Pump from surface water (Surface water Pump)	124	64.58	27	46.55	151	60.40
(4) Pump from groundwater (GW Pump)	24	12.50	3	5.17	27	10.80
Total	192	100.00	58	100.00	250	100.00

Source: Field survey (2022)
[a]The four FLI types will be designated as Gravity, Manual, Surface water Pump and Groundwater Pump

A total of 250 irrigators, made up of 192 men and 58 women, were involved in the survey. Ten key informant interviews were also done with community leaders, personnel of the District Departments of Agriculture (DDA) and the Ghana Irrigation Development Authority (GIDA). Table 5.1 gives the types of FLI and the numbers of men and women respondents. Men are generally more involved in the FLI than women particularly at the production level, even though it was found that in the Tempane District, there were more women irrigators than men.

5.3 Results and Discussions

5.3.1 Socio-Economic Characteristics of Irrigators and Irrigated Farming Systems

Table 5.1 above gives interesting information with regards to irrigation types commonly undertaken by men and women. Most of the farmer-led irrigation in the study area is by Surface water Pump (60.4%). Figure 5.2 displays surface water pump and shallow well bucket irrigation system in one of the study areas. Of the 60.4% Surface water Pump irrigators, men are 49.6% while women constitute 10.8%. Considering all the farmer-led irrigators interviewed in the study area, 59.2% of them who use pumps are men and only 12.0% are women. Also, among men irrigators, over 77% use pumps while among women

irrigators about 51% use pumps. The results clearly indicate the disadvantaged position of women in the use of pumps in farmer-led irrigation. It shows that the claim by Dittoh et al. (2013) that smallholder "irrigators (in the Upper Region) generally regard pumps as saviours" may not be true of women irrigators. Women in the Bawku West District stated that, "women find it difficult to use pumping machines even though they are aware that they are more efficient and can help them to irrigate larger areas". They think that technology is for men. The women find it more convenient to use buckets to fetch water to irrigate manually. The men in the Bawku West District agree that the use of pumps for irrigation is inconvenient for women. Women and men in the other districts stated different reasons why women may be disadvantaged in the use of pumps. The reasons include high cost of pumps, difficulties in operating them, cost of fuel, dangers to them since the petrol pumps must be carried home every day etc. Women and men in the Tempane District disagreed with most of the reasons given in the other districts for the limited use of pumps by women. They assert that the main constraint is the non-availability and high cost of the pumps. They want women to be financially supported in the purchase of pumps. That is to indicate that there is always a need to obtain localized information on gender and local level issues. The women in the Tempane District had been practicing irrigated agriculture for long and that may be the reason they have adequate knowledge and experience in the use of pumps.

Fig. 5.2 Surface water pump irrigation (L) and shallow well bucket-fetch irrigation (R)

Table 5.2 Basic socio-economic characteristics of men and women farmer-led irrigators

Characteristics	Men		Women	
	Mean	Standard deviation	Mean	Standard deviation
Age (years)	42.62	9.83	43.81	8.43
Number of years of schooling	5.39	5.56	4.10	4.92
Rainfed farmland size (ha)	2.25	1.16	1.84	1.18
Irrigated land size (ha)	0.78	0.35	0.64	0.41

Source: Field survey (2022)

Table 5.2 gives basic socio-economic information of the men and women irrigators. The average ages of the men and women irrigators are about the same with the women's average age being slightly higher. Even though in most household and farmer surveys, men farmers tend to be older, it may not be so with respect to irrigated agriculture because younger women have less chances of being able to secure irrigated plots compared to older women. The table also shows that men have larger rainfed farmlands and irrigated lands than the women. The differences are however not as wide as reported in similar studies in West Africa (Akuriba et al. 2010). The gender dynamics in agriculture in the northern part of Ghana have been changing quite significantly due to many factors, including migration and climate change effects (Vercillo 2021). Women tend to take on more roles in agricultural production as young men in the northern parts of the country migrate to the south in search of better living conditions. Even though women from the north also migrate for similar reasons, more young men compared to young women migrate.

Tables 5.3 and 5.4 give the various kinds of crops and crop mixtures cultivated by the men and women irrigators. They are the typical irrigated farming systems that exist in the study area. Even though some of the crops, such as onions, okra and pepper are presented as sole crops there are usually "minor" crops such as local leafy vegetables, sweet potatoes etc. mixed in the plots (basins). Mixed (inter) cropping tends to be the norm. An analysis of the decision by both men and women to cultivate particular crops reveals the importance each gender attaches to the different crops. The reasons for those decisions should be important to decision makers and other stakeholders involved in farmer-led irrigation development. The two tables indicate that vegetable cultivation is the focus of farmer-led irrigation in the Upper East Region of Ghana. They are regarded as high value crops and thus farmer-led irrigation is clearly for commercial purposes despite the small sizes of the irrigated plots. Staple cereal crops are not irrigated at all in the area. Very few farmers however cultivate beans, a common grain legume but mainly for the leaves as a vegetable.

Onions, okra and pepper, either as sole crops or in mixtures are the crops cultivated most by men by Gravity, Manual and Groundwater Pumps

Table 5.3 Men irrigated farming systems (crop/mixtures) by irrigation types

Crop/crop mixtures			Crop/crop mixtures		
Gravity	**Freq.**	**% Freq.**	**Groundwater pump**	**Freq.**	**% Freq.**
Onion	6	3.13	Pepper	10	5.21
Okra	4	2.08	Onion/pepper	5	2.60
Onion/pepper	3	1.56	Onions	3	1.56
Leafy vegetables (LV)/Okra	2	1.04	Pepper/Leafy Vegetables (LV)/okra	1	0.52
Pepper	2	1.04	Pepper/LV	1	0.52
LV	1	0.52	Tomato/pepper	1	0.52
Cabbage/okra/garden eggs	1	0.52	Onions/tomato/ pepper	1	0.52
			Onions/tomato/pepper/LV	1	0.52
			Onion/pepper/okra	1	0.52
Manual	**Freq.**	**% Freq.**	**Surface water pump**	**Freq.**	**% Freq.**
Onions	14	7.29	Onion	31	16.15
Onion/okra	2	1.04	Pepper	13	6.77
Onion/pepper	2	1.04	Tomato	10	5.21
Tomato	1	0.52	Onion/pepper	7	3.65
Onion/tomato	1	0.52	Tomato/pepper	4	2.08
Onion/LV	1	0.52	Onion/okra	3	1.56
Pepper/okra	1	0.52	Pepper/LV	2	1.04
Onion/pepper/LV	1	0.52	Pepper/okra	2	1.04
Tomato/LV	1	0.52	Onion/LV/okra	2	1.04
Tomato/onion/LV	1	0.52	Onion/beans	2	1.04
			Tomato/LV	2	1.04
			Onion/tomato/cabbage	2	1.04
			Onion/pepper/okra	1	0.52
			Onion/LV/okra	1	0.52
			Onion/cabbage/LV	1	0.52
			Onion/calabash	1	0.52
			Cabbage/okra	1	0.52
			Pepper/green maize/beans	1	0.52
			Other crop mixtures	14	7.28

Source: Field survey (2022)

as given in Table 5.3. Tomato production is also important in Surface water Pump irrigation. In the case of the women in the study area, leafy vegetables, okra, onions and pepper are most preferred and are also cultivated as sole crops or in mixtures (Table 5.4). According to both the men and women, the decision to cultivate a crop as sole or in mixtures is based on many factors that cannot be easily enumerated. The main objective is however to reduce risks. They could be risks related to production, marketing, water availability, labour availability and several others.

Comparison of Tables 5.3 and 5.4 shows the many kinds of crop/crop mixtures (irrigated farming systems) the men cultivate compared to the women. The main reason is the larger irri-

gated areas the men have. Also, most men in the study area start irrigation in October/November while most of the women start in January/February. The last quarter of the year (October to December) is the period for harvesting of rainfed crops and women spend much longer time doing the harvesting (Fig. 5.3) and farm level processing of rainfed crops.

The number of times an irrigated crop appears as sole or in mixtures can be used to infer the relative importance of the crops to the farmers. The results of that computation for both men and women are given in Table 5.5. The table shows that while onions, pepper and tomato are the top three irrigated crops for men; onions, okra and leafy vegetables are the top three crops for

Table 5.4 Women irrigated farming systems (crop/crop mixtures) by irrigation types

Crops/ crop mixtures			Crops/crop mixtures		
Gravity	**Freq.**	**% Freq.**	**Groundwater pump**	**Freq.**	**% Freq.**
Leafy vegetables/Okra	3	5.17	Pepper/ LV	1	1.72
Leafy vegetables	2	3.45	LV	1	1.72
Okra	2	3.45	onion/okra	1	1.72
Onions	1	1.72			
Manual	**Freq.**	**% Freq.**	**Surface water Pump**	**Freq.**	**% Freq.**
Onion	9	15.52	Onions	9	15.52
Okra	4	6.90	Pepper	4	6.90
Leafy vegetables /okra	3	5.17	Onion/pepper/beans	2	3.45
Onion/tomato	2	3.45	Onion/pepper	2	3.45
LV/tomato	1	1.72	LV/Okra	2	3.45
LV	1	1.72	onion/LV	2	3.45
Tomato	1	1.72	Okra	1	1.72
			Calabash	1	1.72
			Tomato	1	1.72
			Okra/garden egg	1	1.72
			Tomato/pepper	1	1.72

Source: Field survey (2022)

Fig. 5.3 Harvesting of irrigated produce: onions (L) and leafy vegetables (R). women are largely into the harvesting and marketing components of the irrigation value chain

women. That is quite revealing. Even though the main aim of irrigated production is commercial, the women still lean towards irrigated crops that are consumed almost daily by households in the communities. The leafy vegetables are the local ones. It is instructive that while about 4% of men cultivated cabbage (an exotic vegetable), none of the women cultivated it as sole or as a mixture. Hope et al. (2009) reported that cabbage is not a common crop for women given the resource (labour, agro inputs) intensive nature of cabbage production. Several decades have passed since that finding and it seems the situation has not changed.

The strong point of FLI in the Upper East Region is its continued production of indigenous (agroecological) crops that have the potential of contributing significantly to nutrition security and food sovereignty. The production processes should therefore also now move back to agroecological practices by concentrating on the use of local production inputs rather than imported fertilizers and agrochemicals. That will not only free farmers from the high costs of inputs but will ensure the production of safe and nutritious food. That is not to imply that there is no need for integration of appropriate modern irrigation practices to FLI. There is a strong need to integrate with

Table 5.5 Relative importance of irrigated crops – men and women[a]

Crops	Men		Women	
	Frequency	% frequency	Frequency	% frequency
Onions	127	66.15	28	48.28
Pepper	68	35.42	10	17.24
Tomato	30	15.63	6	10.34
Okra	26	13.54	17	29.31
Leafy Vegetables	22	11.46	16	27.59
Cabbage	7	3.65	0	0.00
Garden egg	3	1.56	1	1.72
Calabash	3	1.56	1	1.72
Beans	3	1.56	2	3.45
Green maize	2	1.04	0	0.00

Source: Field survey (2022)
[a]Computed from Tables 5.3 and 5.4. The frequences are obtained by counting the crops whether sole or in mixtures in the tables

Table 5.6 Harvested values (in US$/ha) of irrigated produce by irrigation type and by gender

FLI irrigation type	Men		Women	
	Harvested value (in US$/Ha)	Standard deviation	Harvested value (in US$/Ha)	Standard deviation
(1) Gravity flow from surface water	2675.96	2103.18	570.62	696.26
(2) Manual (bucket-fetch) from surface water and/or groundwater (wells)	1122.43	975.60	518.97	826.62
(3) Surface water Pump	1929.13	1690.89	1638.66	1666.98
(4) Groundwater Pump	3333.30	2538.77	1630.50	2209.40
Total	2074.37	1883.92	1086.95	1810.37

Source: Field survey (2022)

5.3.2 Harvested Values (Gross Margins) of Irrigated Produce

The complexities in the irrigated farming systems with regards to farm sizes and the crop/crop mixtures makes it difficult to determine yields of the various crops to any appreciable degree of accuracy. However, since a large proportion of the produce is sold, the value of the harvested produce can be estimated more accurately. Table 5.6 gives the computed values of the harvested produce per hectare. It should be noted that the average area is about 0.8 hectares for the men and about 0.6 for the women. That means most of the irrigators did not earn up to the amounts stated in

Table 5.6. Also, the information in the table are gross margins and may not reflect profits earned. The irrigators, however, indicated that they are very cautious in the use of expensive inputs, and they largely believe the gross margins reflect the relative profitability of the different irrigation types.

Table 5.6 shows that, for men, the irrigation type with the highest gross margin is Groundwater Pump, followed by Gravity flow, and Manual has the least gross margin. For the women, the irrigation type with the highest gross margin is the Surface water Pump, followed closely by the Groundwater Pump, and Manual again has the least gross margin. In all cases, the standard deviations are high indicating wide variations in the harvested values (gross margins). That is to be expected given the way the irrigated crops are marketed. There are not many producers who have targeted buyers of the produce so the prices

improved marketing, storage, transportation and processing knowledge and processes. FLID should be aiming at appropriate integration.

can vary very widely daily or even within the same day. The standard deviations with respect to the harvested values of the women is very worrying. All of them are higher than the mean values indicating that variations are larger than the mean values. There are high risks in FLI, and government intervention is necessary to ensure sustainability of the system. Most of irrigators are beginning to see a difficult future for FLI without government intervention.

Figure 5.4 shows pictorially that the harvested values (gross margins) of the produce of the men irrigators are higher than those of the women for all the irrigation types. Also, the Gravity and Groundwater Pump are those that disadvantage the women the most, but Surface water Pump is much more women friendly. The women in most of the areas agreed that the Gravity and Manual irrigation requires much more labour than Surface water Pump and Groundwater Pump.

Reasons for the seeming inefficiency of production by women as compared to men, as given in Table 5.6 and Fig. 5.4, were sought from both genders during follow-up visits to the communities. The women said the reasons for the situation are numerous. They, for example, pointed out the heavy workload they carry during the period, especially at the beginning of the irrigation season which is October to December as stated earlier. According to the women the harvesting of rainfed crops is very labour demanding in addition, the real issue with starting irrigation early is to take advantage of the late rains as well as to sell the irrigated produce at much higher prices. Produce from early irrigated production commands high prices due to low supply. Most of the women, however, say they do not have problems with it because they work with their husbands as families, so they all benefit from the arrangements. Another reason for low output from women irrigated farms is with respect to the use of inputs. The men are much more able to buy production inputs such as improved seeds and fertilizers. Most of the women depend on local seeds and use of manure and compost, which is often not readily available. They, however, say that the use of local seeds and manure have benefits. Irrigated produce from such farms, according to them, are safer, healthier, more nutritious and do not spoil fast. Many of the men agreed that irrigated produce from local inputs is much better in several ways than what is produced using fertilizers and other synthetic inputs. The overall problem however is how to obtain quality manure and compost in reasonably good quantities.

The major issues that both men and women irrigators brought out as challenges that need resolution to give a future to FLI include: (a) decreasing land available for irrigation, (b)

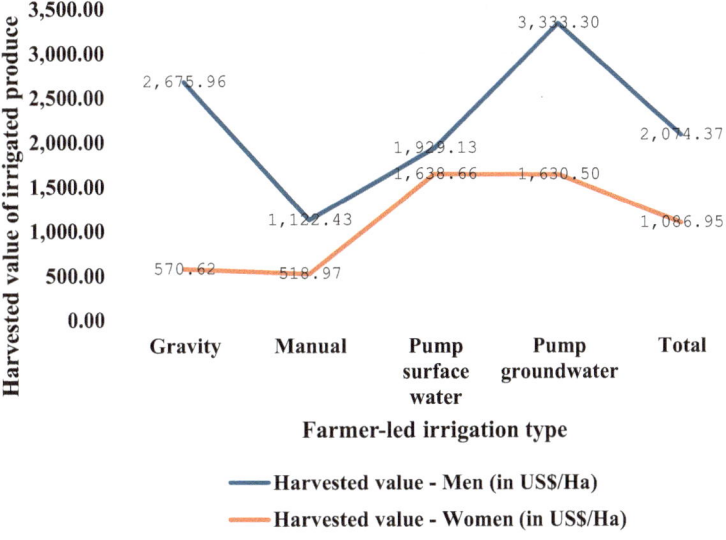

Fig. 5.4 Harvested values (in US$/ha) of irrigated produce by irrigation type and by gender Source: Field survey (2022)

decreasing surface and ground water availability, (c) increasing or high cost of production inputs especially fertilizers and agrochemicals, (d) lack of finance to purchase pumps and other equipment required in irrigated agriculture, (e) lack of ready market and adequate storage facilities for the easily perishable products, and (f) lack of government attention to FLI. They noted that though they all face these challenges they are more severe for women. A key informant (a man) said that "in all the challenges faced, women will never get any attention until the men are catered for. Women can only benefit from the overflow, unless they are specifically targeted". Women in one of the FGDs opined that women empowerment is a deception because they only attend reoccurring meetings with no concrete actions.

The situation of the FLI in Ghana is one where the people are not having the opportunity to integrate their indigenous irrigation knowledge with an appropriate "plug-in" knowledge. It seems development practitioners do not know what improved irrigation technology can plug into the FLI. That is unfortunate because all the listed challenges have examples of solutions from other jurisdictions. Thorough discussions of the many intersecting and cross-cutting issues at various levels should be the beginning of arriving at what can be done. Several ad hoc solutions have been attempted without success. Development projects have come with irrigation components, targeting small scale irrigation but have not achieved much. Examples include, Northern Rural Growth Project (NRGP), Ghana Commercial Agricultural Project (GCAP) and others.

5.4   Conclusion

There is urgent need to increase the production of all types of food if the sustainable development goals are to be achieved, even partially. The current situation of food production under rainfed conditions given the prevailing climate change conditions is very disturbing. In Ghana, and particularly in the Upper East Region, the effects of climate change have continued to be severe.

Droughts and floods have become frequent occurrences. Irrigated agriculture, to supplement rainfed production, is a necessity and the type of irrigation that has proved to have the potential of success is farmer-led irrigation and therefore its development should be prioritized. FLID should, however be done appropriately by continuing to rely on the ingenuity of local irrigators (the plug-in principle). Considerable information is now known about FLI but the issue of inclusivity with respect to active women participation is yet to be well understood. This study has added to the knowledge required for addressing challenges that confront women in FLI by identifying areas that need specific attention. Appropriate agroecological practices in irrigation will free women from the perennial issue of lack of resources to purchase expensive imported inputs. Water lifting devices that are convenient for women are still unavailable. The alternative of investing in drip irrigation also faces initial high investment costs and the problems of lack of appropriate funding mechanisms. Given the growing importance of women in irrigation in the Upper East Region, there is need for the women farmer groups to be more active and dynamic in advocating for greater attention to FLID, with emphasis on women empowerment in irrigated agriculture.

References

Akuriba MA, Dittoh S, Issaka BY, Bhattarai M (2010) Women farmers' perspectives on micro-irrigation technologies in the West African Sahel. In: Paper presented at the 3rd Conference of the African Association of Agricultural Economists (AAAE). South Africa, Cape Town

Anayah F, Kaluarachchi JJ, Pavelic P, Smakhtin V (2013) Predicting groundwater recharge in Ghana by estimating evapotranspiration. Water Int 38(4):408–432

AU (African Union) 2013. The Maputo Commitments and the 2014 African Union Year of Agriculture. Retrieved from 131008_ONE_Maputo_FINAL.pdf

Ayo AA (2020) Water resource investment, financing and development of an agricultural system. In: Proceedings of the 2nd International Conference on Irrigation and Agricultural Development (IRAD 2020). WACWISA, UDS, Tamale

Dietz AJ, Millar D, Dittoh S, Obeng F, Ofori-Sarpong (2004) Climate and livelihood change in North-East

Ghana. In: Dietz AJ, Ruben R, Verhagen A (eds) The Impact of Climate Change on Drylands with Focus on West Africa. Kluwer Academic Publishers, Dordrecht. Chapter 12, pp 149–172

Dittoh S (1991) Efficiency of Agricultural Production in Small- and Medium-Scale Irrigation in Nigeria. In: Doss CR, Olson C (eds) Issues in African Rural Development 1991. Winrock International Institute for Agricultural Development, Arlington. Chapter 8, pp 152–174

Dittoh S (2020) Assessment of Farmer-led Irrigation Development in Ghana. World Bank, Washington DC. www.worldbank.org/gwsp

Dittoh S, Awuni JA, Akuriba MA (2013) Small pumps and the poor: a field survey in the Upper East Region, Ghana. *Water Int* 38(4):449–464

FAO, IFAD, UNICEF, WFP and WHO (2022) The State of Food Security and Nutrition in the World 2022. Repurposing food and agricultural policies to make healthy diets more affordable. FAO, Rome

Giordano M, De Fraiture C, Weight E, Van Der Bliek J (2012) Water for wealth and food security Supporting farmer-driven investments in agricultural water management. IWMI, Accra. http://awm-solutions.iwmi.org

Hope L, Cofie O, Keraita B, Drechsel P (2009) Gender and urban agriculture: the case of Accra, Ghana. In: Hovorka A, de Zeeuw H, Njenga M (eds) Women feeding cities mainstreaming gender in urban agriculture and food security. Practical Action Publishing, pp 65–78

Lefore N, Giordano M, Ringler C, Barron J (2019) Viewpoint-sustainable and equitable growth in sub-saharan Africa: what will it take. Water Alternatives 12(1):155–168

Mohan PC (2002) Nigeria: the National Fadama Development Project. Africa Region Findings & Good Practice Infobriefs. World Bank, Washington, DC, p 83

Niang I, Ruppel OC, Abdrabo MA, Ama E, Christopher L, Jonathan P, Penny U (2014) Africa. In: Barros VR, Field CB, Dokken DJ et al (eds) Climate change 2014: impacts, adaptation, and vulnerability. Part B: Regional aspects. Contribution of working group II to the fifith assessment report of the intergovernmental panel on climate change. Cambridge University Press, pp 1199–1265

Namara RE, Horowitz L, Nyamadi B, Barry B, Strategy G, Program S (2011) Irrigation Development in Ghana: past experiences, emerging opportunities, and future directions. IWMI, Accra. http://www.ifpri.org/themes/gssp/gssp.htm

Obuobi E, Ofori D, Agodzo SK, Okrah C (2013) Groundwater potential for dry-season irrigation in North-Eastern Ghana. Water Int 38(4):433–448

Ofosu EA (2011) Sustainable irrigation development in the White Volta sub-basin: UNESCO-IHE PhD Thesis. CRC Press

Ogunjimi LAO, Adekalu KO (2002) Problems and constraints of small-scale irrigation (Fadama) in Nigeria. Food Rev Intl 18(4):295–304. https://doi.org/10.1081/FRI-120016207

UN 2016. The Sustainable Development Goal Report. Available at: https://unstats.un.org/sdgs/report/2016/The%20Sustainable%20Development%20Goals%20Report%202016.pdf

Vercillo S (2021) A feminist political ecology of farm resource entitlements in Northern Ghana. Gender, Place & Culture 29(10):1467–1496

Woodhouse P, Veldwisch GJ, Venot JP, Brockington D, Komakech H, Manjichi Â (2017) African farmer-led irrigation development: re-framing agricultural policy and investment? J Peasant Stud 44(1):213–233. https://doi.org/10.1080/03066150.2016.1219719

Challenges to Food Systems' Knowledge Integration at Community Levels – Ghana, Kenya and South Africa

Efficacy of Traditional Dispute Resolution Mechanisms in Facilitating Peacebuilding Between Herdsmen and Farmers in the West Mamprusi Municipal of Ghana

John Peter Okoro, Saa Dittoh, and Dzigbodi Adzo Doke

Abstract

Farmer-herder conflict is gradually dominating violence conflict discourse, particularly in West Africa. In Ghana, the conflict is menacing the livelihoods of rural dwellers that are known to be dependent on small-scale crop farming and livestock rearing. The Peace Council of Ghana sees farmer-herder conflict as one of the country's three most significant threats to peace. It is also a major threat to the attainment of food and nutrition security and SDG2 in the country. Relying on qualitative primary data and aligning with the analytical anchorage of hybrid peace governance theory, this Chapter examines the effectiveness of traditional dispute resolution mechanisms (TDRMs) in facilitating peacebuilding between herdsmen and farmers in the West Mamprusi Municipal of Ghana. Utilising both purposive and snowball samplings, 11 focus group discussions (involving 95 participants) and 33 key informants' interviews were conducted in 10 communities in the West Mamprusi Municipality to achieve the research objectives. Findings indicate that the TDRMs and their application for herder-farmer conflict management and peacebuilding in the communities are solely in the hands and determination of the local chiefs. This means that the chiefs directly settle conflict cases between herders and farmers. They, however, sometimes form committees that help to take decisions on the cases. In the past, the TDRMs worked well. Both farmers and herders, however, now perceive the mediation processes at the chiefs' palaces to be corrupt, authoritative, and non-participatory. The respondents trace the problem to the way TDRMs have been influenced by "modernized" chieftaincy and local political systems, which are highly monetized, and which abhors hybrid peace governance systems where all stakeholders must be involved. We argue that the TDRM system should retrace its steps and ensure that salient aspects of the indigenous methods which ensured that all stakeholders are well represented in the TDRMs is adhered to. Fulani herders' elders, community elders, farmers' association leaders or representatives, religious leaders, etc. are required to be part of the system to make it work meaningfully and successfully.

J. P. Okoro (✉) · S. Dittoh · D. A. Doke
University for Development Studies, Tamale, Ghana
e-mail: joo12317@uds.edu.gh

© The Author(s) 2025
S. Dittoh et al. (eds.), *Integrating Indigenous and Scientific Knowledge for Sustainable Food Systems in Africa*, Sustainable Development Goals Series,
https://doi.org/10.1007/978-3-031-85512-2_6

Keywords

Peacebuilding · Farmer-herder conflict ·
Traditional Dispute Resolution Mechanisms
(TDRMs) · Institutional Arrangements ·
Hybrid Peace Governance theory · West
Mamprusi Municipal · Ghana

6.1   Introduction

In West Africa, escalation of violence due to disagreements, mainly between crop farmers and
cattle herders, has become a significant challenge
to peacebuilding. In this region, conflicts between
sedentary farmers and nomadic or semi-nomadic
herders typically result from rivalries over grazing lands, water sources, and other resources.
Population expansion, environmental degradation, climate change, racial tensions, poor governance, and ineffective land and resource
management laws are the most blamed among
the many reasons that frequently contribute to
these disputes. While rapid population growth
increases strain on land and water resources,
environmental deterioration and climate change
decrease the availability of grazing grounds and
water resources, affecting traditional migration
patterns (Brunnschweiler and Bulte 2009; Moritz
2010; Bukari 2016; Cabot 2016; Walwa 2019).
According to Chukwuma (2020), the impact of
climatic conditions, population growth, overgrazing and mechanised farming, and the lack of reg-
ulatory policies to manage the association
between land users, are some of the root causes
of the conflict. All these factors combined affect
farmers and herders, hence the struggle to secure
their livelihoods. Several West African nations
have been embroiled in severe farmer-herder
clashes which have resulted to serious humanitarian challenges including food and nutrition
insecurity.

For instance, in Nigeria (the main hotspots of
the conflict in West Africa), massacre between
September 2001 and May 2004 accounted for
more than 2000 deaths (IRIN, 2004 in Cabot
2016) and the escape of 20,000 Fulani nomads
(Moritz 2006). This series of violent clashes was

between Christian Tarok farmers and Muslim
Fulani herders (IRIN 2004) and is said to have
started with cattle theft (Moritz, 2006). Moreover,
farmer-herder clashes are responsible for the loss
of thousands of lives between 1990 and 2015 (Uji
2016), while costing the nation an annual loss of
about US\$14 billion (Ogundipe and Josiah 2016).
Approximately 1229 and 2000 deaths linked to
the conflict were reported in 2014 and 2015
respectively (Amzat et al. 2016). Also, from
January 2016 to June 2018 (30 months), attacks
due to farmer-herder conflict occurred 95 times,
claiming 2272 lives, which further indicates that
at least, 3 lives are lost every day due to the conflict in Nigeria (Okoro 2018). In addition, thousands   of   people   from   different   agrarian
communities across Nigeria, particularly in the
north central region of the country have been displaced and confined to camps owing to the conflict (Uji 2016; Okoro 2020). These confrontations
have been marked by disagreements over ownership of land, cattle destroying crops, and issues of
ethnicity and religion.

In Burkina Faso, the increased crossing of
border by the nomadic herders in search of grazing land for their livestock is a source of worry to
the government, which thinks that such activity
may lead to cross-border conflicts between farmers and herders (IRIN, 2010). However, it is estimated that 60% of Burkina Faso's central-southern
herders live on the other side of the border, that
is, in Ghana (IRIN 2010). Cabot (2016) referring
to IRIN's (2010) report, revealed that 18 deaths
and many unevaluated injured persons due to
farmer-herder confrontations took place between
2007 and 2010 in the southern part of the country.
In 2009 alone, 29 cases of land damage by livestock were registered in the southern province of
Nahouri, Burkina Faso (IRIN 2010).

Farmer-herder disputes in some West African
nations like Mali, Burkina Faso, and Niger have
been made worse by the presence of armed
groups, notably jihadist organisations. These
organisations take advantage of existing tensions,
escalating disputes between sedentary farming
communities and nomadic herders, leading to
widespread displacement, and creating an unstable environment (Krätli and Toulmin 2020). In

north and central Mali, terrorists groups linked to AI Qaeda in the Islamic Maghreb and the Katibat of Macina are taking advantage of the farmer-herder situation to recruit fighters (Krätli and Toulmin 2020). Many Peul (Fulani) feel politically and economically marginalised, which provides fertile ground for recruitment by jihadist groups. Increasing numbers of Peul are being compelled into sympathy with the jihadi movement to protect themselves against bandits and other criminal gangs who steal their cattle (Krätli and Toulmin 2020).

In Ghana, like in other West African countries, the farmer-herder conflict is both overt and covert in nature depending on the level of first victims' reactions in a place. It is overt when the conflict involves observable violence – when it arouses open confrontation between the two groups which may lead to injury, killings and/or arson. It is, however, covert when the conflict is lacking observable violence – a situation in which confrontations may occur but secretly settled either through a common understanding between the two groups or a third party intervention to achieve a win-win outcome. For example, a farmer may realise that his/her farm has been trespassed on or destroyed by herders' cattle and decide to approach the herders in person or report to the third party to seek for compensation. Once it is settled amicably without violence, such conflict can be said to be covert. Ethnically, again like in other West African countries, the Fulani tribe is the most dominant pastoral group in Ghana (Abbass 2014). As a result, the farmer-herder conflict in Ghana implicates Fulani herdsmen and mostly Ghanaian settled farming communities. According to Tonah (2002), the 1920s and 1930s are regarded as the period the Fulani herders (along with their cattle) from Burkina Faso, Mali, and Niger began migrating into Ghana for economic reasons such as better pasturelands and water to graze their cattle. After settling in Ghana, the herders soon discovered a new opportunity – tending cattle for Ghanaians for a fee (Tonah 2004). In other words, Ghanaian livestock owners started hiring the services of the Fulani herders to tend their livestock. Be that as it may, local farmers were said to be in cordial relationships

with the herders due to mutual exchange opportunities occasioned by the cattle by-products and the after-harvest crops foliage associate to the herder and farmers respectively. There was a kind of symbiotic relationship between the herders and farming communities.

Today, the presence of Fulani herdsmen is conspicuous in almost all Ghana's agro-ecological zones (Appiah-Boateng 2020), with both their own cattle and those belonging to Ghanaians. Thus, menaces emanating from open grazing are becoming as conspicuous as the presence of the herders and the cattle almost across the length and breadth of the country. Most often, land encroachments, crop destruction, depletion of economic trees, etc. are among these menaces of which the aftermaths are usually confrontations, violent clashes, attacks and counter attacks which would further lead to deaths, loss of properties/livelihoods, and displacements (Alimba 2014; Okoro 2020). One of the cases in Ghana included the killing of about 80 cattle owned by the Fulani herders in Dumso in the then Brong-Ahafo Region, which was carried out in February 2016 by enraged youth as a reprisal to the destruction of farmers' crops in the area (Stanley, Dary, Harvey, and Asaah 2017). In essence, the farmer-herder conflict situations create outcomes that hinder socio-economic improvements and livelihoods of the people (Nchi 2013; Okoro 2018). Moreover, researchers have demonstrated that the conflict does not only affect individuals' socio-economic wellbeing but also suffocates the food system, by creating food insecurity (Antwi 2018; Oji and Anih 2021; Nnaji et al. 2022).

The work of Krätli and Toulmin (2020), *Farmer-herder conflict in sub-Saharan Africa*, published by the International Institute for Environment and Development (IIED) revealed that from 1997 to 2013, Asante Akim North District in Ashanti Region and Northern Region were the major hotspots of farmer-herder conflict. The work also revealed that locations such as the New Juaben, Afram Plains, and Kwahu East in the Eastern Region, Asante Akyem North and Afigya Sekyere in the Ashanti Region, Jirapa-Lambussie in the Upper West Region, Pru in the Brong Ahafo Region, and West Mamprusi in the

Northern Region were the major hotspots of the conflict from 2013 to 2017. In addition, Olaniyan (2015), Bukari (2016), Ahmed and Kuusaana (2021) identified Gushegu District in the Northern Region as one of the hotspots of the conflict in Ghana, considering the August 2011 reprisal attack on the Fulani herders in which 14 Fulani herders including their families were killed and some of their houses razed. Less attention has, however, been paid on the farmer-herder conflict case of West Mamprusi Municipal. West Mamprusi Municipal in the North-East Region was thus selected for this study.

Successive governments in Ghana have made efforts to solve the farmer-herder problem through different policy approaches. Unfortunately, the problem has continued to escalate. Noting the ineffectiveness of existing policy efforts, there is a need for a more comprehensive study of the problem to unearth changes that have occurred over time, using an intensive community-based and hybrid (integrated) peace governance approach. The approach calls for exploration of the role of traditional institutions and other recognized bodies in the management of farmer-herder conflict and how their weaknesses and strengths can inform future policy in Ghana. This is because local and community-based dispute resolution mechanisms have been shown to be effective in mitigating farmer-herder conflict and aiding post-conflict recovery in other jurisdictions (Crisis Group 2018). Some traditional dispute resolution mechanisms (TDRMs) in local peacebuilding, particularly within some African settings, have proven to be all-inclusive and guided by the principle of leaving no one behind in dialogue (Ajayi and Buhari 2014). Those TDRMs encourage forums that promote beneficial coexistence between farmers and pastoralists while managing potential threats. They aim to remove the root causes of conflict and reconcile conflicting parties genuinely. The argument here is that some TDRM tend to work well, and even in the study area farmer-herder conflicts were resolved amicably in the past. This study was to establish the robustness or otherwise of this argument, using the situation in West Mamprusi as a case study.

Based on the foregoing, this Chapter examines the efficacy of traditional dispute resolution mechanisms (TDRMs) in facilitating peacebuilding between farmers and herdsmen in the West Mamprusi Municipal of Ghana. The study from which the Chapter is derived sought to answer the following questions:

1. What TDRMs exists for managing farmer-herder conflict and peacebuilding in communities in the West Mamprusi Municipality?
2. What are the farmer-herder conflict management processes in the communities?
3. How do the people perceive the mediation processes at the local chiefs' palaces in the communities?
4. How effective is the TDRMs in managing the farmer-herder conflict?

6.2   Relevant Literature

6.2.1   Farmer-Herder Conflict

Farmer-herder or herdsmen-farmers conflict is a natural resource-based conflict occurring between cattle herders and crop farmers. In essence, farmer-herder conflict is the systematic aggression exhibited by nomadic pastoralists in the context of socio-ecological struggles with sedentary farmers (Okoli and Atelhe 2014). Crop destruction, rape of female farmers inside the farm, serious bodily harm (including amputation of body parts), killing of farmers (of different genders) blamed on the Fulani herdsmen, cattle rustling, killing/shooting of cattle by crop farmers, rejection of Fulani herdsmen as members of communities where they were born and raised to adulthood, and killing of Fulani herdsmen including their families (women and children) are among the dangers associated with farmer-herder conflict (Alimba 2014; Opoku 2014; Baidoo 2014; Bukari and Schareika 2015; Appiah-Boateng 2020; Ahmed 2022; Ugwueze et al. 2022).

6.2.2 Traditional Dispute Resolution Mechanisms (TDRMs)

The TDRMs concept of peace is comprehensive and broader than the Western conception of absence of violence, covering both negative and positive peace, intra-personal peace, interpersonal peace, within communities, with nature, and peace with God (Gena and Jarra 2023). Traditional dispute resolution generally refers to the methods and practices used in societies or cultures that have longstanding customs and norms for addressing disputes. These practices vary across different cultures and regions. In most societies, TDRMs takes the form of alternative dispute resolution (ADR) approach which is known in the Western court system, introduced when cases demands negotiation, mediation, and conciliation mechanisms (Mcquoid-Mason 2021). Thus, the importance of TDRMs in negotiating and mediating peaceful settlements between conflict parties cannot be overstated, and as a result can be applied in the case of farmer-herder conflict. For instance, Brottem (2021), citing the Kabara dispute resolution committee in Adamawa State of Nigeria that has contributed to reducing farmer-herder conflict in the state, averred that TDRMs when properly instituted and/or adopted can yield a robust result in the management of farmer-herder conflict.

Scholars have underlined the importance of TDRM processes globally over time, noting that courts only handle a small portion of all disputes that occur in society (Galanter 1981; Muigua 2018). In fact, some African countries (such as Kenya) have ingrained it in their constitution. For instance, article 159 of Kenya's 2010 Constitution made provision for TDRM as alternative to formal court system, and mandates courts and tribunals to nurture it (Muigua 2018). African cultures have their own methods for resolving disputes before colonisation. The parties to a dispute might negotiate whenever one occurs. In other situations, the Council of Elders or senior citizens could serve as neutral parties to mediate the issue. Additionally, elders and close family members might mediate disputes amicably and counsel disputants on the importance of living in harmony. Therefore, traditional methods of resolving disputes were designed to promote peaceful coexistence among Africans (Muigua 2018). Also, there were some institutions, values, and traditions that were essential in resolving disputes. Traditional institutions of conflict resolution were well-structured in cessation of hostilities, peaceful conflict resolution, strengthening social relations, ensuring stability, promoting harmony, and healing families in the past (Gena and Jarra 2023).

TDRMs have contributed to de-escalation of violent conflict and averting malevolence and played essential roles in promoting peace and stability in resource-based conflict communities (Gena and Jarra 2023). The aim of these methods is to make everybody involved in the resolved conflict happy and be at peace with each other again. These mechanisms can play significant roles in building a culture of peace in any society. Mechanisms for resolving disputes under customary law can be very helpful in lowering the amount of litigation in both criminal and civil matters. These techniques include bargaining (negotiation), conciliation, adjudication, mediation, and reconciliation (Otieno 2016). The first three are typical of Western concepts of ADR designed to cut down on litigation. Traditional courts and arbitration mechanisms, which place a strong focus on rapprochement (reconciliation) and reintegration, are, however, not typical of ADRs (Mcquoid-Mason 2021). Traditional dispute resolution processes tend to adopt a "communal" rather than the Western "individualistic" approach to dispute settlements (Mcquoid-Mason 2021). In traditional societies great emphasis is placed on reconciliation and reintegrating the disputing parties back into their communities. The purpose of traditional dispute resolution processes is to not only reconcile the relationships between individuals, but also the relationships of the individuals with their communities. Dispute resolution mechanisms also enabled the disputants to express themselves fully, without complexity or formality, and yet assured them of a knowledgeable and just resolution that would maintain communal relations (Aiyedun and

Ordor 2016). According to Muigua (2018), TDRM has been successful in areas where it has been used to resolve conflict because it is more accessible to the public, flexible, voluntary, and economical and encourages connections. Therefore, there is an urgent need for a comprehensive strategy to restore the traditional conflict resolution mechanisms and promote cooperative peacebuilding efforts.

6.3 Peacebuilding

Peacebuilding is the development of constructive personal, group, and political relationships across ethnic, religious, class, national, and racial boundaries. It aims to resolve injustice in nonviolent ways and to transform the structural conditions that generate deadly conflict. Peacebuilding can include conflict prevention; conflict management; conflict resolution and transformation, and post-conflict reconciliation. It is typically a much longer-term process that involves a great number and variety of stakeholders, starting with the citizens of the locations in which conflict is about to occur, ongoing, or has ended. According to Galtung (1976), peacebuilding aims to promote sustainable peace by addressing the root causes of violent conflict and supporting indigenous capacities for peace management and conflict resolution. Galtung's work aroused debate on the definition of peace and its approaches such as peacekeeping, peacemaking, and peacebuilding. He called for the creation of peacebuilding structures to stimulate sustainable peace (United Nations 2010). Galtung (1964) described peace research as research into the conditions for moving closer to peace or at least not drifting closer to violence. He stressed that peace is classified into positive and negative peace, defining negative peace as the absence of violence and/or war, and positive peace as the integration of human society or absence of structural violence (indirect violence) (Grewal 2003).

In the context of farmer-herder conflict, peacebuilding efforts should not only concentrate on achieving negative peace (absence of violence conflict) but also on the positive peace (capacity building, institution building, and policies for conflict prevention and livelihood enhancement). Thus, local peacebuilding arrangements in which the local communities that are directly embroiled in the farmer-herder conflict could be integrated in the peacebuilding processes can be helpful. According to Adjei and Hancock (2021), local peacebuilding activities tend to focus on local communities and issues of local governance, which contrast with state-building activities such as security sector reform, institutional reform, constitution writing, and the construction of national democratic structures. Local peacebuilding examines the microlevel impacts of state-building as well as peacebuilding activities that develop at the local level, either for local benefit or as attempts to influence national-level policies and debates (Adjei and Hancock 2021).

6.4 Fulani Herders' Settlement and Relationship with the Traditional Leaders in Ghana

Land ownership and control are exercised by those sitting on the numerous chieftaincy stools or skins in several Ghanaian tribes. In the Northern Region of Ghana land (particularly unoccupied land) is vested in the chiefs (Kasanga 1988; Asaaga 2021). In all parts of Ghana there are laid down customary practices with regards land acquisition and/or land use. In the Northern Region the practice is for a stranger who needs land and wants to dwell there, must obtain permission from the chief and his elders and/or landowners. The same is true for Fulani pastoralists who want to establish in a specific region. They typically seek information about settlement from other pastoralists, livestock traders, and intermediaries in the livestock business. A pastoralist would advise other Fulani residing in the settlement or the livestock merchant about their plan to settle in the area after they are convinced that the resources of the area are favorable and sufficient for cattle herding. The established Fulani would speak with the Chief and Landowners on the new immigrant's behalf (Rabbe 1998). The chief and elders would allot the new immigrant a plot of land on which to settle in exchange for agreed payments which could be in cash or kind as well as agreement to comply with necessary traditional procedures and regulations. It is typical for two or more pastoralists to give animals to landowners in exchange for rent to settle in the area and utilize the local pasture and water resources. Most of the time, the chiefs, landowners, stock owners, etc. that invite the Fulani herders to their community immediately benefit from their presence. In exchange for being permitted to settle on the landowners' property and utilize neighboring land for cultivation and pasture, the herders pay these landowners rent in cash and in kind. According to a 2009 police report, chiefs and landowners in these rural areas derived significant income from rent payments made by migrating Fulani herdsmen (Bonye et al. 2021). Chiefs prefer to provide land to the migrating Fulani, particularly the herdsmen who are wealthy in livestock and can afford to pay significant sums as settlement fees (Bonye et al. 2021).

6.5 Hybrid Peace Governance Theory (HPG)

A Western liberal peace paradigm's intrusion and the complexity of contemporary conflicts are just two of the significant obstacles that traditional African conflict resolution and peacebuilding mechanisms have had to overcome recently (Murithi 2008; Paalo 2022). However, conventional liberal peace strategies have mostly failed to advance lasting peace and have even deteriorated security conditions in some contexts, according to critical peace study (Mac Ginty 2015; Mac Ginty and Richmond 2013). Therefore, it is argued that achieving sustainable peace largely depends on the participation of local norms and power structures in international peacebuilding efforts, giving rise to the concept of a "hybrid peace" (Boege 2018).

Hybrid political orders, according to Paalo and Issifu (2021), encourage the use of traditional resources and knowledge, foster local ownership and legitimacy, and so assure long-term stability. This claim is supported by a large body of empirical research, such as that conducted in Nigeria (Brottem 2021), Ethiopia (Gena and Jarra 2023), Somaliland (Boege et al. 2009), and Cambodia (Lee 2019). These studies show that hybrid peace has been promoted in these places and has attracted local participation and a sense of ownership of the peace processes and outcomes. Thus, theory of hybrid peace governance (HPG) connects the interaction and collaboration between formal and informal peace governance systems in addressing complex societal challenges. It recognises that in many contexts, including Africa, traditional and customary institutions coexist alongside formal state institutions. The theory of hybrid peace governance can be applied to the study of African traditional institutions and farmer-herder conflict management in the following ways:

1. Coexistence of Formal and Informal Institutions: it acknowledges that African traditional institutions often operate alongside formal state institutions. In the context of farmer-herder conflict, it identifies the relevance of traditional institutions in resolving disputes and managing conflicts, alongside formal legal and governance structures, and how TDRM interact with and adapt to formal legal frameworks, policies, and practices.

2. Hybridisation and Blending of Norms and Practices: HPG emphasises that formal and informal institutions can influence each other and blend their norms, practices, and decision-making processes.

3. Complementarity and Coordination: HPG recognises that effective governance often requires coordination and collaboration between formal and informal institutions. The complementarity between traditional institutions and formal institutions in conflict resolution and peacebuilding efforts as well as exploring opportunities for coordination, information-sharing, and joint decision-making for effective conflicts management.

4. Legitimacy and Trust: HPG explores the legitimacy and trust that communities place in traditional institutions, often based on cultural

and historical foundations. Understanding the role of traditional institutions in building trust and fostering community acceptance can inform strategies for conflict management and peacebuilding that are rooted in local contexts and community norms.

Our submission pushes the Hybrid Governance Theory further to argue that for sustainable and resilient peace building the formal (conventional) liberal peace strategies should "plug-in" to the traditional (local) peace strategies in the way described by Dittoh (Chap. 1). The aim of the blending (integration) of the 'informal' and 'formal' must be to "better" what is existing and not to replace it. This is because as explained above unadulterated TDRMs have been largely successful in a number of jurisdictions.

6.6 Methodology

6.6.1 Research Location

The Republic of Ghana is a decentralised democracy made up of 16 regions, which are further divided into districts, local councils, and unit committees. The research was conducted in the West Mamprusi Municipality (WMMA) of the North-East Region. Walewale is the capial of the municipality and is located between Tamale (the capital of Northern Region) and Bolgatanga (the capital of the Upper East Region. Latitudes 9°55′N and 10°35′N and longitudes 0°35′W and 1°45′W are the boundaries of the West Mamprusi Municipality and it has a total land area of 2610.44 km² and a population of 175,755 according to the 2021 population and housing census (GSS 2022).

Livestock production is one of the main occupations of the people in the North-East Region. A market called 'Walewale Cattle Market' was established in 2022 in the Municipal by the Assembly to boost livestock production in the area (Fugu 2022). The market is aimed at promoting increase in the production of high-quality cattle and generate employment possibilities. The Fulani community has been optimistic that the

project will significantly enhance the lives of the populace, especially Fulani herdsmen, by boosting the cattle trade (Fugu 2022). One may wonder why upon all these development and integration efforts, the Municipal is noted as one of the hotspots of farmer-herder conflict in Ghana. There is the need to seek intensive peacebuilding arrangements to contain the conflict.

6.6.2 Research Design

A qualitative methodological approach was deployed for this study. This is based on the interpretative nature of the research objectives. According to Matthew and Ross (2010), interpretivist epistemology prioritises people subjective interpretations and understandings of social phenomena and their own actions. Also the qualitative approach allows opportunity to obtain rich and more detailed data that capture description of people's stories, feelings, opinions, and beliefs (Matthew and Ross 2010). In addition, the data required to solve the research puzzle in question is unstructured in nature, constructed by the people in their own way. The notion is that by using this technique, the actors in the farmer-herder dispute in West Mamprusi Municipality will be comprehended from within their subjective experiences (Todres and Holloway 2006).

6.6.3 Sampling Method, Sample Size, and Method of Data Analysis

Purposive and snowball sampling techniques were used to select communities and respondents for study. The West Mamprusi Municipality was purposefully chosen after meeting a number of criteria, including (a) that it is one of the hotspots for farmer-herder conflict in northern Ghana (Krätli and Toulmin 2020); and (b) that it is understudied in comparison to other farmer-herder conflict hotspots in Ghana (such as Agogo in the Asante Akyem North District and Gushegu District in the Northern Region.). To select the communities, personnel of Government

Table 6.1 Types of FGDs in communities

Name of community	FGD group	Number of participants
Kurugu	Farmers (male and female adults)	9
Kurugu	Herders (male youth)	10
Wungu	Farmers (male adults)	8
Kparigu	Farmers (male adult)	11
Logre	Herders (male youth & adults)	7
Wulugu	Farmers (male youth)	8
Janga	Farmers (male youth & adults)	9
Kperiga	Herders (male adults)	10
Kukua	Farmers (female adults)	6
Nayiri-Fong	Herders (male youth)	9
Fogni	Farmers (female adults)	8
Total		**95**

Source: Field study (2023)

Table 6.2 Key informant interviews (KIIs)

Category of respondents	Targeted groups/persons	Number of KIIs
NGOs personnel	Community development and field workers	3
West Mamprusi District Assembly personnel	4 Departments (1 person from each of	4
Farmer Associations	Leaders	3
Herder Associations	Leaders	2
Community members (from 10 communities)	Chiefs, Assemblymen, farmers, herders and opinion leaders	21
Total		**33**

Source: Field study (2023)

departments in the West Mamprusi Municipal Assembly (WMMA) assisted in identifying communities that had been most impacted by farmer-herder conflict. Nine (9) communities identified as the municipality's primary conflict hotspots and 1 additional community with moderate cases were selected. The 10 communities are: Kurugu, Wungu, Kparigu, Logre, Wulugu, Janga, Kperigu, Kukua, Nayiri-Fong and Fogni. Mainly focus group discussions (FGDs) were conducted in the communities (see Table 6.1). The same Municipal Assembly personnel helped to identify knowledgeable people with respect to farmer-herder conflicts for key informant interviews (KIIs) (see Table 6.2).

A total of 95 people took part in eleven (11) FGDs that were held in the communities. In total, 128 participants (95 for the FGDs and 33 for the KIIs) participated in the discussions and interviews. The discussions and interviews concentrated in obtaining information on the farmer-herder conflict management process at the local level in the West Mamprusi Municipality and how the conflict parties perceive the mediation processes at the chiefs palaces, as well as to know the effectiveness of TDRM in managing the conflict. The discussions and interview guides were structured in a way that the people were allowed to relax and own the moment to explain the issues the way they felt and what they thought would be solutions to the problem. Thematic and content analytical methods were used to analyse the information obtained. Since the data collection involved interaction (including physical contacts) with humans, ethical procedures such assurance of confidentiality, consent, etc. were observed.

6.7　Analysis and Results

6.7.1　Perceptions of the Traditional Dispute Resolution Mechanism for the Management of Farmer-Herder Conflict in the West Mamprusi Municipal

6.7.1.1　Perceptions of Personnel of WMMA and Its Departments

To understand the existing TDRM for the management of farmer-herder conflict in the West Mamprusi, we asked personnel of government departments[1] at the West Mamprusi Municipal Assembly of their knowledge about established local dispute resolution mechanism targeted at facilitating peace between farmers and herders in the conflict-prone communities. We also asked questions on the people who directly play roles to contain clashes, quarrels, or any form of misunderstanding between the farmers and the Fulani herdsmen in the area. In response, we were informed that the conflict is managed at the community levels and that local chiefs are exclusively responsible for most of the land, water, and tree resources-based conflicts, including farmer-herder conflict. Thus, the farmer-herder conflict in the area is solely managed by the local chiefs with less interference from those outside the royal palaces, except when the conflict involves violent confrontations in which the local police get involved. One of the officials of the Department of Agriculture in the municipal interviewed, stated that:

All arrangements for farmer-herder conflict resolution and management are made at the chief palaces because most of the disputes involving lands and trees are settled by the chiefs. For us at the Department of Agriculture, because we deal directly with the farmers, we only intervene by helping the farmers whose crops are destroyed by cattle to estimate the cost of crops destroyed. Once they get the estimates, then they present them to the

chiefs during mediation, reconciliation, and adjudication processes at the chief palaces.

An interview with the Municipal Chief Executive (MCE) of the WMMA indicates that the municipal assembly only tries to intervene through the Municipal Security Council (MUSEC) when the conflict is not resolved at the community level, or when the conflict constitutes a threat to peace in any of the communities within the municipal. However, the MCE mentioned that:

Sometimes, we go to the communities to assess the issues based on people's complaints. We try to collaborate with the chiefs to find sustainable solutions, one of which was to give the herders times they should let the cattle out to graze. It should be when the people are able to identify who is responsible for the destruction of their farms and not in the night when farmers are asleep. Unfortunately, some of them do not comply with the arrangements. But in all, local chiefs are responsible for the settlements of farmer-herder conflict in their communities.

6.7.1.2　Perceptions at Community Levels

The chiefs, farmers, herders, and the cattle owners (of native affinity) were also interrogated to understand how the conflict occurs and how they are being managed at the communities. From the key informant interviews and FGDs, it was confirmed that the chief palaces are the seats of all mediation, reconciliation, and adjudication processes channeled at resolving conflict between crop farmers and cattle herders in the communities. The TDRM is operated in the form of alternative disputes resolution mechanism (ADRM) in which a win-win approach is preferred. The chiefs alluded to making efforts to ensure that winners don't take it all. The chiefs have committees who sit to adjudicate cases between herders and farmers from time to time. They employ more mediation, reconciliation and adjudication methods instead of litigation, even though the information obtained indicated that the chiefs are authoritative, which sometimes hamper the entire mediation process. Most of the time, the chiefs try resolving the conflict by acting solely as the third party without involving others (say the committee members) by inviting the parties as well as the cattle owners and initiating negotia-

---

[1]The personnel were from the Office of the Municipal Chief Executive, the Department of Planning and the Department of Agriculture.

tion. The chiefs negotiate with the parties to ensure that victims of the conflict are compensated while also assuring that the other party is not exploited. Most times, when an agreement is reached on how much the victim is to be compensated, the cattle owners are invited to pay the money instead of the herders. However, some cattle owners try to share the loss with the herders. In the case the cattle are discovered to belong to the herders, they are to pay the money themselves. From the words of some of the local chiefs:

> Here in Kurugu, the victims of conflict report to me and sometimes I invite them when I hear the case. I invite them immediately I hear that there is quarrel or fight between farmers and Fulani herders so that the problem will not degenerate to serious violence. I call the parties to assess the destruction of the farm to enable us solve it at the palace. When I try to solve it fast between the two parties and one of the parties or both refuse to agree on something, then I invite council members who form committee to further negotiate and convince the parties to accept the available terms for the sake of community peace. The farmers are often compensated, but they do not usually get the full amount they demand for.

> Here in Logre, it is usually after germination of crops and early harvesting periods that the animals start to destroy the crops. I do delegate people to assess the loss whenever it occurs and then begin a resolution process by calling both parties to my palace to ensure the victims are compensated. When this process fails, I organize a committee to reinitiate the mediation and adjudication process, I do this to ensure that no one will feel cheated. The committee recommends a final solution which when brought back to me, the parties must accept it, or allow them to go to the local Police to resolve their issues. Going to the Police Service is very rare. We ensure that the farmers or herders are paid their losses depending on who is the victim of the conflict. Sometimes, retaliation happens because people think they don't get justice in the palace, they choose to take law into their hands. During the resolution process, we call owners of the cattle that caused the destruction to take part in the negotiation of how much to compensate the farmers and what payment method to apply and when to pay.

> Here in Wungu, the Fulani (herders) are now natives because we have been with them for a very long time. In terms of voting, they are voting here. The relationship is cordial. The Fulani people are taking care of our cattle. Fulani migrated from where they were and settled here, knowing that their business is tending animals, we give them our cattle to take care of, which they have been doing for a very long time. Those who are taking care of the cattle are born here. They are different from the nomadic Fulani which only bypass with their cattle, and don't stop in this community. So our major problem is the nomadic ones who are destroying farms indiscriminately. For the sedentary Fulani herders, whenever the cattle they tend destroy farms, I call them and the owners of the cattle to compensate the farmer. The farmer normally charge the herders and I help them to collect the money for him/her. We go and look at the farm and estimate the cost of destruction – the number of acres that were destroyed, then we will now charge the herders based on the acres or based on the amount that the farmer spent on the farm. Sometimes, the case may be free. When they destroy small, the farmer will get small.

It is also interesting to note that the chiefs do not only resolve conflict by acting after it has occurred, but also invest preventive efforts (peacebuilding approach), such as conducting meetings with the Municipal Assemblies, the Fulani chiefs, and the farmers (in a kind of hybrid peace governance style) to insure common grounds for peaceful coexistence between the Fulani ethnic groups and the natives. During the rainy season, the chiefs go out to mark demarcations where the herders can use to penetrate the bush safely without encroaching on the farmers' cultivated farmlands. For example, one of the Assemblymen interviewed divulged that:

> During the rainy season, our chief organizes people who go out to create a grazing route and crossing space to prevent the cattle from using people's cultivated farmlands as their crossing routes. This effort is to prevent farmland encroachment and destruction of crops and its aftermath, yet cattle still destroy crops. I don't know why.

In Janga community, the chiefs revealed that:

> To put preventive measure against confrontation occurrences, I had a meeting with some of the natives and told the farmers not to attack the Fulani people whenever they found that cattle have destroyed or are destroying their crops, but to check the farm and estimate the amount lots and/or what was spent on the farm and inform me. But if they wound the cattle, that case is completely lost. It is better to report the matter instead of confronting the cattle or the tenders so that we will find a way to compensate farmers for the destruction.

During a FGD session with the male adult farmers, one of the discussants shared his experience
as follows:

> There is a Fulani chief here. We don't have any
> problem with the Fulani except with their animals.
> I have farms and people who work for me there.
> Sometimes, some of the workers who do not have
> patience to follow the process of obtaining justice
> beat up the Fulani herdsmen in the farm when
> crops are destroyed by the cattle. But I have
> stopped them from doing that. I told them not to
> beat the herdsmen whenever they see the Fulani
> and their cattle in the farm, they should rather
> come and report to me so that I can report the case
> to the chief palace. There is some level of under-
> standing and some progress. When crop farmers
> beat herdsmen when the destroy their farms, they
> cannot ask for compensation. It means they have
> taken the law into their own hands. The herdsmen
> are, however, a headache for us.

With respect to the role of the police in farmer-
herder situation in the Municipal a Fulani Chief
said that:

> Because of the local rules here, you cannot take a
> case to the police station. When you bypass the
> Chief and go to the police station, the case will get
> back to the Chief for settlement. Sometimes, the
> chiefs themselves will go to the police station and
> ask for the cases to be released so that it can be
> settled at the community level. On some rare occa-
> sions, some farmers go to the police station when
> they feel dissatisfied with the settlement process at
> the chief's palace. It is, however, difficult for us
> (herders) to report cases to the Police people even
> when the native people injure us or the cattle.

6.7.1.3 Summary of the TDRM Processes at the Community Level

Conflict usually occurs whenever the cattle
destroy farms or whenever the cattle are stolen or
injured and the culprit is found in the act or at
least when the victim is able to trace the culprit.
The victim approaches the other person for compensation and peaceful settlement. If there is
acceptance and agreement is made, then the case
can be settled without a third party. However,
many victims of the conflict prefer to report to
the community chief. If a case is reported to the
community chief, there are three levels of
ADRMs. Firstly, there is direct negotiation
between the herder, cattle owner and farmer to
determine the damage and necessary compensation. Such negotiations, however, often fail, and
herders usually prefer the second level. At the
second level, a local committee is set up by the
local chief. This committee system is neither statutory nor formalised or registered. Membership
includes representatives from interest groups
such as landlords, youth, cattle owners, and resi-
dent Fulani herders. The committee will listen to
the two parties and arrive at a decision for compensation. The committee usually meets parties
several times depending on the issue. Farmers
frequently accuse committee members of favor-
ing herders due to bribery and corruption and are
frequently dissatisfied with their actions (or lack
thereof). Additionally, the committee lacks the
legal authority to carry out its recommendations,
such as compensating parties or kicking herders
out of the neighborhood. The third level is for the
local chief to take a decision if decision-making
at the committee level is unsuccessful. The
farmer and the herder are expected to abide by
the chief's decisions, which are traditionally
binding. Failure to follow the chief's instructions
may result in expulsion from the community or
other penalties. But much like the committee,
there is frequently a lot of suspicion in the chief's
decision-making. Due to this mistrust, the problem is then turned over to the local police if none
of these local remedies succeed since the parties
involved fear unfair treatment. Since herders are
not Ghanaian citizens, they often feel they will be
unfairly treated at the police stations. The farmers
also are not usually comfortable with the involvement of the police because they feel the herdsmen can bribe their way through since they tend
to be richer. The other advantage the herdsmen
have is that they take care of cattle of several
influential people in the society, and they often
intervene on their behalf (Fig. 6.1).

6.7.1.4 Parties' Perceptions of the Mediation Process at the Chief Palace

The results from the FGDs and KIIs indicate that
the parties in conflict are neither comfortable
with nor trusting the existing TDRM in the communities. When the farmers were asked if they

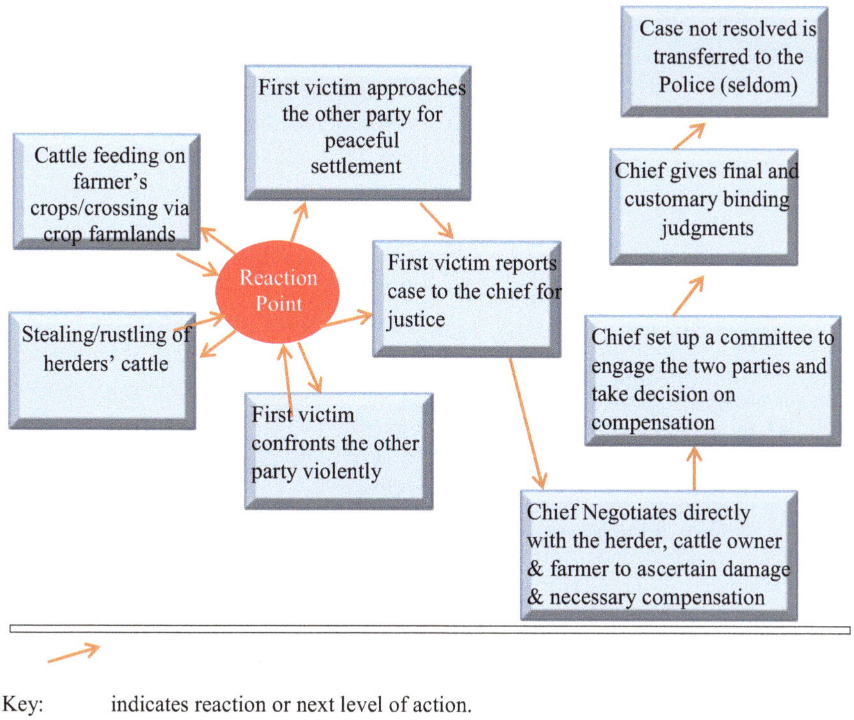

Key: indicates reaction or next level of action.

Fig. 6.1 Conflict management process at the community level in WMM. (Source: Authors' construct based on field data. Key: Indicates reaction or next level of action)

were satisfied with the way the cases were handled at the chiefs palaces, they unanimously alluded to being dissatisfied because the "chiefs always support the foreigners" (Fulani herdsmen) instead of their own people. They see chiefs to be partial, unfair, corrupt, and authoritative in the way they manage the conflict. They also perceive the conflict management process to lack transparency, participation, equity, and justice. These, according to them, are the reasons they prefer not to take cases to the chiefs but to confront the herders for compensation or seize their cattle in exchange to the crops destroyed. They believe that reporting the case to the chiefs is a waste of time since they cannot get any justice. Most of the farmers also believe that the chiefs are always bribed with cattle by the Fulani herders to win cases against the farmers. When asked if they have alternative authorities to direct the case to aside from the chiefs, they mentioned that they go to the police stations on rare cases—

when the case involves death casualty and/or when they feel dissatisfied with the chiefs' judgment. However, many of them revealed that they fear going to the police stations to report a case that chiefs have tried to resolve because of the consequences. They mention that the consequences may include banishing from the community or confiscation of farmers' farmland by the chiefs. For instance, one of the discussants during a FGD with the male youths at Janga community, revealed that:

> We can only report to the Chief whenever the 'Fulani cattle' destroy our farms. Although we do settle with them sometimes without going to the chief when they cooperate. It is unfortunate that cases taken to the chief appear to be a waste of time sometimes because getting justice there is hard, and when you dare disagree with what the chief concludes about the case and try taking it to the Police station, chief may collect your farmland from you or banish you from the community because he is the custodian of the land we live and farm in.

Also, a farmer revealed that:

> When my entire maize farm was destroyed, I traced the cattle legs to a Fulani's home. I went to them to pay me all the damages, but they were ready to fight me. I took the case to the chief but even after officials of Department of Agriculture had estimated the cost, I didn't get up to a quarter of it because, I believe, the 'Fulani' has bribed the chief.

In one of the communities, an opinion leader who was a key informant opined that:

> The people don't trust the Chief. Farmers have been complaining to me that whenever they take a case to the chief, he prefers to collect money or cows from the 'Fulani people' and close the case without ensuring the farmers are compensated for their damages. Even when they want to compensate them, he/she will only get a meagre amount from the whole loss, that's after the case would have lingered for so long.

The same questions were put to the herders. They, the herders, were generally satisfied with the chiefs' methods of settlements; even though many of them, just as the farmers, believe that the mediation and adjudication process is corrupt, authoritative, and non-participatory.

In the word of one of the Fulani chiefs:

> We cannot say that there is unfairness and partiality in the way the settlements are conducted. We are satisfied and okay with it. What you should know is that no matter how good the process is there must be other things such as corruption and show of power. Sometimes when we give money to the chiefs, we don't know if the farmers get it or not because, even after giving the compensation they ask for, they still accuse us and insult us of destroying their things without paying. They also forget that most of the cattle are not ours and we don't deliberately send the cattle to their farms.

Another key informant who preferred to be classified as a 'Fulani Elder' instead of 'Fulani Chief' in one of the communities had this to say:

> If you look round, you will see that all the cattle are stationed at one place until a particular period— morning or evening before we move them around for grazing. This tells you that we are trying our best to ensure that no farm is destroyed, and whenever it mistakenly happens and the owner of the farm reports to the chiefs, we go there and pay. We are satisfied with the chief's adjudication process, but sometimes we don't have the opportunity to negotiate because everything there is about money.

> We also cannot go to the police station because of the ethnic problems we face.

The perception of the farmers and herders on the mediation and adjudication processes inherent in the existing TDRMs in the communities vary with regards to their fairness and satisfaction of the parties. They, however, agree that processes are corrupt, authoritative, and non-participatory in nature. That explains why the farmers feel dissatisfied while the herders feel more satisfied. Herders tend to be more able and ready to pay their way through especially as they are regarded as "foreigners". *There is "monetization" and adulteration of the TDRMs as decried by a key informant.*

6.7.2 Effectiveness of TDRM in Managing Farmer-Herder Conflict and Peacebuilding in the West Mamprusi Communities

Despite the information already revealed by the participants about the TDRM process available in the community, we probed further by asking their perception about the effectiveness of the TDRM in general. Some of the information from the key informant interviews indicate that the conflict resolution arrangements by the local chiefs, though not good enough, have been responsible for non-escalation of the conflict to monumental violence in the communities. There is however fear that the murmuring, distrust, anger, and frustration that are being expressed could lead to aggression and/or trigger violence if unattended to.

Most participants are of the view that the TDRM is not effective and needs more of negotiation than authoritarianism. An interview with an Assemblyman from one of the communities indicates that the existing TDRM is not effective because:

> Even though some of the Fulani herders in the communities have been 'enskinned' (made chiefs) by the efforts of the native chiefs through the national government integration policy, the cordiality between the Fulani ethnic group and the

natives is yet to be felt. That alone is indication of the ineffectiveness of conflict resolution mechanisms that are in place.

In the same vain, all the farmers who participated in the FGD in all the communities asserted that the TDRM is not effective. Some of the herders are also of the view that the TDRM needs improvement. The herders frequently mentioned that, "usually there is not much negotiation, you are to accept anything the chiefs say; and the worse is that the farmers over estimate their losses".

Generally, the TDRM is not effective due to the following factors:

1. Lack of proper negotiations and renegotiations between parties during the TDR process.
2. The TDRM is solely managed by the chiefs and not with elders and other community stakeholders such as religious leaders, opinion leaders, Assemblymen, etc.
3. The TDRM is not institutionalized in the communities; instead, it is only an arrangement by the chiefs of which they involve only those they wish to during the mediation and adjudication process. It is far from being an effective hybrid peace governance system.
4. There is age-long distrust of the land resource custodians by the users. Example, the farmers may never trust the chiefs' intervention in any case involving them and the herders.

6.8 Policy Recommendations

1. The research findings indicate that a well-structured TDRM can be effective in managing farmer-herder conflict. A major constraint in the Ghana case is the ethnicity divide problem, The Fulani herdsmen who have been living in Ghana for decades, some of whom their grandfathers were born and nurtured in Ghana, who already do not have other countries to call their own, need to be integrated into the communities as indigenes. In this way, both the natives and migrants would see themselves as one people and can constitute

elements to be involved in the TDRM arrangements together.
2. Hybrid peace governance which is a bottom-up approach in peacemaking processes that originate from the local people themselves, when utilised would guarantee local ownership of peace process (Issifu and Bukari 2022). Although the approach operates through customary institutions, local chiefs alone cannot personalise the process inherent in the institutionalisation and operation of the institution. The personalisation is the main reason behind the mistrust that exists in the communities, as all efforts of the chiefs are not recongnised by the people. Therefore, implanting a proper TDRM that allows collaborations with the relevant stakeholders, and would report to appropriate national institutions in a situation where cases are difficult to handle at the local level or where negotiations reach impasse and the parties are unsatisfied, can terminate conflict and build sustainable peace. It is akin to insisting that there should be good understanding and appreciation of the local peace building process before plugging-in others. It is the way to obtain a workable, sustainable and resilient TDRM.
3. To ensure the effectiveness of TDRM, national and local authorities should establish such mechanisms in the various agrarian communities across the nation. These mechanisms should involve various groups, including farmers, pastoralists, community vigilantes (or community police assistants), and state security agencies, to foster inclusivity and trust in the mediation process. Since civil society groups and non-governmental organisations play important roles in promoting dialogue, collaboration with them will be helpful. Like other alternative dispute resolution (ADR) mechanisms, TDRM is expected to be constituted in a way stakeholders are well represented. For instance, the elders of Fulani herders, community elders, farmers' association leaders or representatives, religious leaders, etc. should be part of the system to make it meaningful.

4. A comprehensive institutional arrangement strategy that combines conflict resolution techniques, bottom-up approach of which hybrid peace governance is anchored, resource management reforms, the strengthening of governance structures, investments in agriculture and rural development, and the encouragement of communication and understanding between various communities is needed to address farmer-herder conflicts not only in West Mamprusi, but in Ghana as a whole and in West Africa. Municipal and district authorities in collaboration with the national government can complement each other in managing farmer-herder conflict and preventing violence by collaborating with the chiefs and other community leaders to adopt a suitable TDRM.

5. Finally, since farmer-herder conflict cannot be effectively resolved when lodged in a court of law due to the difficulty in addressing the underlying causes, recognising the role of traditional dispute resolution mechanisms in the constitution will be germane. This is because local peace dialogues, negotiations and reconciliation meetings often result to peace and harmonious co-existence.

References

Abbass IM (2014) No Retreat No Surrender Conflict for Survival between Fulani Pastoralists and Farmers in Northern Nigeria. European Scientific Journal, ESJ, 8, 331–346.

Adjei M, Hancock LE (2021) Local Peacebuilding. In: Richmond O, Visoka G (eds) The Palgrave encyclopedia of peace and conflict studies. Palgrave Macmillan, Cham

Ahmed A, Kuusaana ED (2021) Cattle Ranching and Farmer-herder Conflicts in sub-Saharan Africa: Exploring the Conditions for Successes and Failures in Northern Ghana, African Security, https://doi.org/10.1080/19392206.2021.1955496

Ahmed A (2022) Farmer-herder conflicts in northern Ghana amid climate change: causes and policy responses. Policy Brief 09 DEC 2022. MEGATREND Afrika.

Aiyedun A Ordor A (2016) Integrating the Traditional with the Contemporary in Dispute Resolution in Africa. LAW, DEMOCRACY & DEVELOPMENT/

VOL 20 (2016). P154. ISSN: 2077-4907 https://doi.org/10.4314/ldd.v20i1.8

Ajayi A, Buhari L (2014) Methods of conflict resolution in African Traditional Society. Afr Res Rev 48:138–157. https://doi.org/10.4314/AFRREV.V8I2.9

Alimba NC (2014) Probing the dynamic of communal conflict in Northern Nigeria. African Research Review, 8(1):177–204

Amzat A, Fagbemi A, Lawal I, Wantu J, Akingboye O (2016). Menace of Fulani herdsmen. The Guardian News. Retrieved from https://guardian.ng/features/menace-offulani-herdsmen/

Antwi S (2018) Farmer-herder conflict and food security in Kwahu East District, Eastern Region, Ghana. In: Master's Thesis submitted to the Department of International Environment and Development Studies, Noragric, Norway

Appiah-Boateng S (2020) Land-Use Conflicts and Psychosocial Well-being: A Study of Farmer-Herder Confict in Asente Akym North District of Ghana. University of Cape Coast.

Asaaga FA (2021) Building on "Traditional" land dispute resolution mechanisms in rural Ghana: adaptive or anachronistic? Land 10(2):143. https://doi.org/10.3390/land10020143

Baidoo I (2014) Farmer-herder conflicts: A case study of Fulani herdsmen and farmers in the Agogo traditional area of the Ashanti Region (Doctoral dissertation, University of Ghana)

Boege V, Anne Brown M, Clements KP (2009) Hybrid political orders, not fragile states. Peace Rev 21(1):13–21

Boege V (2018) The international-local interface in peacebuilding. In: Kwesi Aning M, Brown A, Boege V, Hunt CT (eds) Exploring peace formation: security and justice in post-colonial states. Routledge, New York, pp 174–190

Bonye SZ, Aasoglenang TA, Der FB, Bobie CN, Dery G (2021) Fulani herder-farmer conflicts in rural Ghana: perspectives of communities in the Sawla-Tuna-Kalba District. J Plan Land Manag 2(1):77–86

Brottem L (2021) The growing complexity of farmer-herder conflict in West and Central Africa. In: Africa Security Brief A publication of the Africa center for strategic studies No 39l July 2021

Brunnschweiler CN, Bulte EH (2009) Natural resources and violent conflict: resource abundance, dependence, and the onset of civil wars. Oxf Econ Pap 61(4):651–674. https://doi.org/10.1093/oep/gpp024

Bukari KN (2016) Farmer-herder relations in Ghana: interplay of environmental change, conflict, cooperation and social networks (Georg-August University of Göttingen, 2016)

Bukari KN, Schareika N (2015) Stereotypes, prejudices and exclusion of Fulani pastoralists in Ghana. Pastoralism: Research, Policy and Practice, 5(1). https://doi.org/10.1186/s13570-015-0043-8.

Cabot C (2016) Climate change and farmer-herder conflicts in West Africa. In: Climate change, security risks and conflict reduction in Africa. Springer, Berlin, Heidelberg, pp 11–44

Chukwuma KH (2020) *Constructing the Herder–Farmer Conflict as (in)Security in Nigeria*. Afr Secur:1–23. https://doi.org/10.1080/19392206.2020.1732703

Dary SK, Harvey SJ, Asaah SM (2017) Triggers of Farmer-Herder Conflicts in Ghana: A Non-Parametric Analysis of Stakeholders' Perspectives. Sustainable Agriculture Research; Vol. 6, No. 2; 2017 ISSN 1927-050X E-ISSN 1927-0518 Published by Canadian Center of Science and Education. https://doi.org/10.5539/sar.v6n2p141

Fugu, M (2022) Walewale Cattle Market inaugurated. GRAPHIC ONLINE. https://www.graphic.com.gh/news/generalnews/walewale-cattle-market-inaugurated.html

Galanter M (1981) Justice in many rooms: courts, private ordering and indigenous law. J Legal Pluralism 19:3

Galtung J (1976) Three approaches to peace: peacekeeping, peacemaking, and peacebuilding. In: Peace, war and defense: essays in peace research, vol II. Christian Ejlers, Copenhagen, pp 297–298

Galtung J (1964) An editorial. J Peace Res 1(1):1–4

Gena AM, Jarra KI (2023) An appraisal of the practice of indigenous conflict resolution mechanisms in building a culture of peace in Bale zones, Oromia National Region State, Ethiopia. Heliyon 9:e14970

Grewal BS (2003) Johan Galtung: positive and negative peace. School of Social Science, Auckland University of Technology

International Crisis Group. (2018) "Stopping Nigeria's Spiralling Farmer-Herder Violence." https://d2071andvip0wj.cloudfront.net/262-stopping-nigerias-spirallingfarmer-herder-violence.pdf.

Issifu AK, Bukari KN (2022) (Re)thinking homegrown peace mechanisms for the resolution of conflicts in Northern Ghana. Confl Secur Dev 22(2):221–242. https://doi.org/10.1080/14678802.2022.2059934

IRIN (2004) Plateau state violence claim 53,000 lives. Report

IRIN (2010) Cross-border land conflict risks. UN Office for the Coordination of Humanitarian Affairs

Kasanga K (1988) Land tenure and the development dialogue: the myth concerning communal landholding in Ghana. University of Cambridge, Department of Land Economy

Krätli S, Toulmin C (2020) Farmer-herder conflict in sub-Saharan Africa? International Institute for Environment and Development (IIED, 106.

Lee SY (2019) Local ownership in Asian Peacebuilding: rethinking peace and conflict studies. Springer, New York

Mac Ginty R, Richmond O (2013) The local turn in peace building: a critical agenda for peace. Third World Q 34(5):763–783

Mac Ginty R (2015) Where is the local? Critical localism and peace-building. Third World Q 36(5):840–856

Matthews B Ross L (2010) Research Methods. Pearson Longman, London.

McQuoid-Mason D (2021) "Could traditional dispute resolution mechanisms be the solution in post-colonial developing countries – particularly in Africa?", Oñati Socio-Legal Series, 11(2), pp. 590–604. https://doi.org/10.35295/osls.iisl/0000-0000-0000-1145.

Moritz M (2006) Changing Contexts and Dynamics of Farmer-Herder Conflicts Across West Africa Canadian Journal of African Studies / Revue canadienne des études africaines 40(1):1–40 https://doi.org/10.1080/00083968.2006.10751334

Moritz M (2010) Understanding herder-farmer conflicts in West Africa: outline of a processual approach. Human Org 69(2):138–148. https://doi.org/10.17730/humo.69.2.aq85k02453w83363

Muigua K. (2018). Traditional Dispute Resolution Mechanisms under Article 159 of the Constitution of Kenya 2010. . http://kmco.co.ke/wp-content/uploads/2018/08/Paper-on-Article-159-Traditional-Dispute-Resolution-Mechanisms-FINAL.pdf

Murithi T (2008) African indigenous and endogenous approaches to peace and conflict resolution. In: Francis DJ (ed) Peace and Conflict in Africa. Zed Books, London and New, York, pp 16–30

Nchi SI (2013) Religion and politics in Nigeria: The constitutional issues. Jos: Greenworld

Nnaji A, Ma W, Ratna N, Renwick A (2022) Farmer-herder conflicts and food insecurity: evidence from rural Nigeria. Agric Resour Econ Rev 51:391–421. https://doi.org/10.1017/age.2022.9

Ogundipe S, Josiah O (2016) Nigeria loses $14 billion annually to herdsmen-farmers clashes. Premium Times. https://www.premiumtimesng.com

Oji RO, Anih J (2021) Addressing food security challenges in Nigeria. Int J Innov Develop Policy Studies 9(3):163–176

Okoli AC, Atelhe G (2014) Nomads against natives: a political ecology of farmer/herder conflicts in Nasarawa State, Nigeria. Am Int J Contemp Res 4.2:76–88

Okoro JP (2018) Herdsmen/farmers conflict and its effects on socio-economic development in Nigeria. Journal of Peace, Security, and Development, 4(1):143–158.

Okoro JP (2020) Herdsmen – Farmers' Conflict: Implication on National Development (Nigeria in Perspective). International Journal of Scientific & Engineering Research 11(2):808 ISSN 2229–5518.

Olaniyan A (2015) The Fulani–Konkomba conflict and management strategy in Gushiegu, Ghana. J Appl Secur Res 10(3):330–340. https://doi.org/10.1080/19361610.2015.1038763

Opoku P (2014) Exploring the causes and management of pastoralists-farmer conflicts in Ghana. Journal of Energy and Natural Resource Management (JENRM), 1(3). https://doi.org/10.1007/s13280-016-0805-6

Otieno HJ (2016) The Effectiveness of Traditional Dispute Resolution Mechanism in Facilitating Women's Access to Justice in Land Disputes[online]. LLB Research Dissertation. University of Nairobi, pp. 19–20

Paalo SA, Issifu AK (2021) De-internationalising hybrid peace: state-traditional authority collaboration and conflict resolution in Northern Ghana. J Inter State-Build 3(15):406–424

Paalo SA (2022) Intergenerational gaps in women's grassroots peacebuilding in Ghana: a critique of "inclusive peacebuilding", Journal of Aggression, Conflict and Peace Research, Vol. 14 No. 4, pp. 287–303. https://doi.org/10.1108/JACPR-01-2022-0663

Rabbe J (1998) Aspekte des Viehhandels im Norden Ghanas unter besonderer Berücksichtigung der Beziehung zwischen Frafra-Kleinhändlern und Fulbe-Viehzüchtern

Todres L, Holloway I (2006) Phenomenological research. The research process in nursing, 5, 23–35

Tonah S (2002) "Fulani Pastoralists, Indigenous Farmers and the Contest for Land in Northern Ghana," Africa Spectrum 37, no. 1: 43–59. https://doi.org/10.2307/40174917

Tonah S (2004) Defying the Nayiri: Traditional Authority, People's Power and the Politics of Chieftaincy Succession in Mamprugu/Northern Ghana. Legon Journal of Sociology, 1, 1: 42–58.

Ugwueze MI, Omenma JT, Okwueze FO (2022) Land-Related Conflicts and the Nature of Government Responses in Africa: The Case of Farmer-Herder Crises in Nigeria. Society. Springer Nature https://doi.org/10.1007/s12115-022-00685-0

Uji WT (2016) Catholic church diplomacy: the challenge of Fulani herdsmen terrorism in central Nigeria. Online J Arts Manag Soc Sci 1(1):52–60

United Nations (2010) UN Peacebuilding: an orientation, Peacebuilding Support Office

Walwa WJ (2019) Growing farmer-herder conflicts in Tanzania: the licenced exclusions of pastoral communities interests over access to resources. J Peasant Stud:1–17. https://doi.org/10.1080/03066150.2019.1602523

Demons of Developments Without Dialoguing with Indigenous Knowledge

Moses Mwangi

Abstract

The genesis of this Chapter deliberations is informed by the fact that both the pre and post independent Kenyan governments have never prioritized people's knowledge to be part of the development agenda. The situation is worse for the pastoralists who live in the arid and semi-arid lands (ASALs). The regions have faced development marginalization with the resultant deleterious effects being heightened by weather whims, unfair policies and scarce resources. For a people who rely on rational exploitation of natural resources for their survival, the effects have led to unsustainable lives and livelihoods. The discussions herein aim at sharing the experiences of the Maasai pastoralists inhabiting the Kajiado County of Kenya whose development interventions from the turn of the century have only known ostracism, with negative short and long term social, economic and environmental consequences. The main premise of the discourse is that the vision of a truly global knowledge partnership will be realized only when development partners participate as both contributors and users of knowledge. There is therefore a need for development support agencies to appreciate the indigenous knowledge of the target localities and plugging in global knowledge for effective and integrated sustainable development of the people and their ecosystems.

Keywords

Marginalization · Commons · Land tenure · Pastoralism · Social capital

7.1 Introduction

In Kenya, development support objectives from without an environment rarely consider indigenous knowledge to play noteworthy roles. The knowledge is often emphasized to lack independence and consequently requiring replacement by other ability systems. The perspectives misses out on the fact that the knowledge of a community can shape a milieu debate and present critical options for development. Local understanding and interpretations have capacity to reflect interests, making it essential for development practitioners to pay attention to indigenous knowledge and for incorporation in development efforts, providing a sense of belonging and an arena which shapes actions of individuals and systems within communities (Brennan et al. 2013). The interactions with the knowledge provides and enable appreciation of local parameters that can

M. Mwangi (✉)
South Eastern Kenya University, Kitui, Kenya
e-mail: mmwangi@seku.ac.ke

© The Author(s) 2025
S. Dittoh et al. (eds.), *Integrating Indigenous and Scientific Knowledge for Sustainable Food Systems in Africa*, Sustainable Development Goals Series,
https://doi.org/10.1007/978-3-031-85512-2_7

help in appreciating sources and magnitudes of local problems and appropriate solutions prescription. The advantage is derived from the fact that indigenous knowledge is rooted in the people's culture which houses diversified values and ideas, experiences in materials and dimensions, rules and regulations.

Consecutive Kenyan governments and development partners' approaches have been sectoral and too broad in scope to adequately take care of unique needs of target areas. The attitude negates the advantage of onboarding existing social and cultural arrangements in order to address interdependencies. Among the Maasai pastoralists, the value of their livelihoods is usually underestimated, contributing to making the people remain marginalized despite their important role in the food and income provision at local and national levels. The community faces numerous challenges that range from the ravages of climate change and land tenure challenges that curtail mobility for seasonal grazing to degradation of natural resources, and unfair public policies. The community reproduction system are based on traditional grazing that use mobility as a key strategy for sustainable management of resources in their environment (Tarayia 2004). The movement enables access to forage and water, increase the digestive efficiency and weight gains of livestock, reduce exposure to food shortages in dry seasons and droughts, and helps livestock avoid disease (Homewood et al. 2009; Tarayia 2004). Customary land tenure based on collective land ownership and reciprocal grazing rights, and allocating specific pastures to different species of livestock have been a norm.

From the late 1990s, indigenous scholars have cried out for integrating indigenous knowledge into discourses and practices (Mihesuha and Wilson, 2004). This is yet to receive total embracement. The experience of the Maasai pastoralists is that of victims of institutional discrimination of their rangeland indigenous knowledge, an aspect that has diminished their roles in their development discourses. The community's traditional resources of resources development and management reveals a deep knowledge in maintaining biodiversity and obtaining maximum

benefits of their ecosystem for climate resiliency, food and nutrition, medicine, and the economy. This chapter is therefore a contribution to community development discussions, with emphasis being on the need for realization of the importance of development endeavors "plugging-in" to indigenous knowledge.

7.2   Study Methodology

The study covered Kajiado County which is located to the south of Kenya (Fig. 7.1). The area is arid and semi-arid that is traditionally home to the Maasai pastoralists. The region has progressively attracted communities from around the globe who have practically changed the rangeland use to all forms of landscape that invite researches to establish the rationale and the knowledge systems applied. This research employed a qualitative research approach under which in-depth interviews, focus group discussions and key informant interviews were carried out. The focus group discussions comprised multi segments of the community including women, men, youth and the elders. For cross triangulation, an in-depth interview was held with individuals considered to be most versed with the Maasai pastoralists and their knowledge in resources development and management. Different cases were investigated and presented to support key findings of the study. Through stakeholder participator approached, six failed projects were identified and the reasons for their expiration followed closely to establish the relationship and indigenous knowledge of the people. The data were contextually and textually analyzed using participatory rapid appraisal tools.

7.3   Conceptual Framework

Indigenous knowledge is appreciated to have multiple symbolizations and different connotations (World Bank 1999). Richards (1985) reputes it to be characterized by attributes of ecological particularism generated in a local natural

Fig. 7.1 Location of Kajiado County in Kenya. (Source: Ndago 2024)

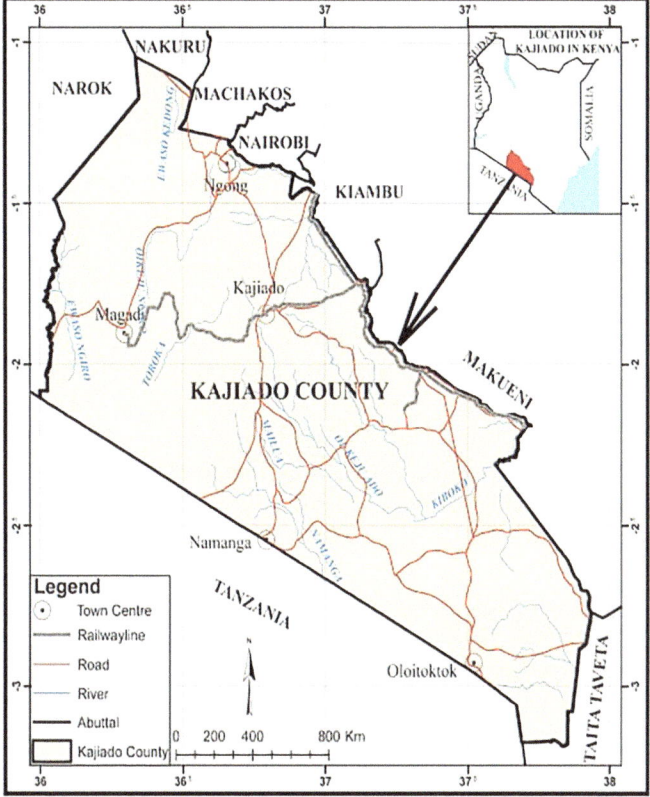

environment and under specific ecological relationships. Warren and Rajasekaran (1993) agrees with the two observations, reckoning indigenous knowledge to provide an information base to facilitate communication and decision making in a particular society. The agents of indigenous knowledge in these contexts are people unique to a given context (Warren et al. 1995). Consensus is built that the applicability of the knowledge is beyond specific locality and, almost the same for communities of one socioeconomic, and political setting. For the Maasai pastoralists, the knowledge plays a critical role in linking the local production strategies with the ecological variability and local institutions. The pastoralists' rangeland sustainable resources development and management depends on organizing capacity of the pastoral systems based on indigenous knowledge whose absence of observance has likelihood to lead to failure in interventions and making the local institutions and the ecosystems more vulnerable (Gadgil et al. 1993). Breman and de Wit

(1983) shares that pastoralists' indigenous knowledge has strength from the basis that it is developed from direct interaction with the people's natural and social environment. Therefore, it subsumes information and skills by which the people can derive the highest benefits from the available resources.

7.4 Maasai Pastoralists Indigenous Knowledge System

The indigenous knowledge of the Maasai pastoralists is ingrained in their culture and depends on what people can reminisce, observe and hear (Kotut and McCrickard 2021) and is usually transmitted through personal communication (Sow and Ranjan 2021). As such the knowledge belongs to the community and, those who possess it have it on behalf of the community (Kereto et al. 2022). The Maasai indigenous knowledge revolves around management of their livestock, water, pasture, drought, animal diseases, natural forestry and, coexistence of livestock with wildlife.

(a) Indigenous Knowledge on Pasture and Livestock Management

The Maasai community uses its indigenous knowledge to raise livestock in the aspects of nutrition, pests and diseases, breeding and, predation. Knowledge in pasture management has the duo purpose of offering prospects of improving the livestock livelihood and preserving natural ecosystems. The association with forage able to distinguish each plant's seasonality, nutritional value, toxicity and medicinal properties for the different animals they keep (Kariuki et al. 2018). The herders take decisions to save their livestock from hunger based on progression and severity of a drought. Rotational grazing is preferred to utilize different parts of the range's plant life in order to maintain healthy and nutritious forages. The knowledge is handy operating mixed herds to take advantage of dissimilar reproduction and feeding habits which ensure survivors during drought as different species are adapted differently (Kereto et al. 2022). Indigenous knowledge customarily demanded harvesting of only dry twigs for firewood while herbal plants received part harvesting to allow appropriate regeneration. Moving from an exhausted grassland area to a new and rich one was traditionally the only way of keeping livestock alive. When livestock returned to previous grazing lands, new edible grass was found. Skinner (2010) endorses that the loosely dispersed, highly mobile populations fair better than sedentary populations in the highly variable environments. This is no longer possible for the Maasai pastoralists under the individual land tenure ownership.

Sometimes, land parcels are paddocked into smaller portions *(Olopololi)* using either permanent or temporary fencing to allow controlled of grazing and conservation of runoff. Persistence of hunger in livestock during drought is met by feeding them on the *acacia tortilis (Oltepesi)* and

acacia nilotica (Olkiloriti) trees. The c*ynodon dactylon* grass *(Emurua)* is administered to upsurge lactation in livestock while the *olea africana (Oloirien*) herb and cider charcoal mixed with milk counters the poisonous effects of *Oldule (ricinus communis)* (Kereto et al. 2022). The pastoralists at times avoid mountainous areas or long-distance walks with livestock to avoid infections and diseases from other livestock as well as abrupt weather changes.

(b) Indigenous Knowledge in Water Development and Management

Under the former communal land tenure ownership, the Maasai pastoralists traversed and made use of watered landscapes that housed pastures. The movement restriction caused by land individualization and consequent blocking of livestock movement paths has placed confinement of grazing and watering (Tarayia 2004). The problem traces origin in the period between 1960 and early 1980s when the World Bank prioritized water interventions among the Kajiado Maasai, and introduced comprehensive changes to encourage a move away from traditional subsistence to a more commercial production. The involvement introduced a number of water-development technologies such as the deep boreholes. The water sources failed over time due to absence of appropriate ownership that resulted from lack of consultation with the local people (ASAL Programme, 1988). Given a choice, the pastoralists would have gone for shallow wells improvement as they understood the structures construction, operation and, maintenance. In the traditional practices of exploring for shallow groundwater, the pastoralists use various plants, specific insects and rocks as water presence indicators. Human divinations knowledge has been positively applied in some instances. The use of ecosystem knowledge, forked sticks (particularly *croton megalocarpus),* brass and steel rods, water-bottle meniscus and pendulums are in application, raising confidence in the practice of exploring for groundwater. The pastoralists observe the height of the thorn tree to estimate the depth of aquifers. The barks of the same tree and the *moringa oloifera* plant are used as water coagulants. Maasai pastoralists' knowledge regarding locating and harvesting geysers has been valuable. Limitations, however arise from a lack of knowledge about how to take the practice to the next level and offer high levels of safety and sustainability.

Indigenous knowledge and practice relevant to the water resources sector include spatial and temporal location, construction, collection and storage, conservation strategies and, management. The water rules and regulations enforcement rely on social ostracism, social rebuke and social isolation. The belief and use of curses is another potentially powerful tool for ensuring adherence to rules. The method is more respected than the modern justice dispensation process. Rules of reciprocal obligation are daily reinforcements of regulations concerning tenure, consumption and the protection of shallow wells and natural resources. Indigenous social controls that include limiting access to certain water resources are however consciously eroded by land-tenure changes, immigration, social disintegration, increasing income gaps, and decreasing resource capacity.

(c) **Indigenous Knowledge in Weather Forecasting**

Extreme weather events have been a recurring challenge to the Maasai pastoralist, leading to employment of a variety of strategies to cope. Varied environmental signs are employed to silhouette endurance decisions that revolve around visible and audible observations. Animal and plants behavior, weather information sharing by travelers and lunar observations are some for the commonest methods of weather forecast. Contours of goats intestines are observed to predict the weather and forth comings in the society, based on the belief that everything done on earth is recorded in the intestines. In recent years, the Maasai pastoralists have adapted oral and visual knowledge sharing through the radio, mobile phones and television. Reflections on collective historical recollections is identified to be offer the pastoral Maasai valuable climatic insights.

(d) **Indigenous Knowledge in Food Security**

The Maasai *ilterito* age group (over 80 of age) talk of food security 'colonization' to imply the loss of indigenous food production and management to the modern food chain. The elders recall that the Maasai have traditionally depended heavily on livestock for nutrition to make a diet largely consisting of milk, meat, fat, blood and honey, laced with different herbs in the food chain. By embracing sedentary life and interacting with communities from without their environment, the Maasai pastoralists' diet has undergone profound changes. However, meat is still an important dietary component of the community that is not only relished for eating, but, is also for multiple social and cultural roles.

Pastoralists have traditionally relied on indigenous knowledge to add value to animal products. Local preservation techniques such as drying, salting, using spices and storing in animal fat are employed to increase the shelf life and, enhancing flavour and taste. As a result, a variety of indigenous meat products have been developed with unique substrates and preparations depending on climate, process and availability of materials. The food processing techniques reflect not only the skills and creativity of the Maasai pastoral people but also their capability to sustain the dynamics of life and ecosystem. To ensure a balanced diet, meat is preserved by mixing it with fat *(olpurda)* and stored in containers for consumption when need arises while milk is stored in calabashes cleaned with the stalk of herbs *(Oloirien)* (Tarayia 2004).

Traditional food techniques were in the past used as a drought coping mechanism, to overcome food insecurity. However, the practices are threatened by the changing lifestyle and reduction of livestock numbers caused by recurring droughts, climate change and effects of land tenure changes. Government and welfare organisations promote production, processing, preservation and marketing of modern foods in the name of diversification. They also support having endogenous plants and intensive land tilling practices, at most not conducive to the bionetwork. The introduction of the new practices has not been keen in taking the people through the entire food chain. The Maasai are increasingly depending on crop products such as cereals, pulses and vegetable oil, purchased from shops or donated by Non-Governmental Organizations who provide relief services or food assistance program especially during drought. This has reduced dependence on livestock products and has had adverse consequences on the knowledge of food preparation methods and processing techniques.

The traditional meat products are reputed to have the potential to provide important levels of key nutrients including monounsaturated fatty acids which are beneficial to health. The knowledge of the product amongst the Maasai community is common, and product are seen to be readily acceptable organoleptically. It is a common practice to add herbs to fresh or boiled milk for various reasons that include flavor, nutrition and medical functions. Some herbs are boiled in water and consumed. Some of the herbs commonly used are attributed to hypolipidemic and antioxidant properties, thus offering one explanation for the low incidence of heart disease despite high consumption of animal-source foods. Almost all herbs are available throughout the year although their availability most likely improves during the wet seasons, and they are generally consumed in a specific season. Wild fruits are consumed mostly by women and children and are highly seasonal, with many increasing availability during and immediately after rainy seasons.

The lessened contribution of traditional diets among the Maasai pastoralists is denying the young in the community indigenous knowledge on food use. The increased drought frequencies spare livestock as sources of food, as migration to areas with pastures is inevitable. Limited income-generating alternatives such as charcoal-making and firewood sales have adverse environmental consequences, ultimately reducing availability of wild foods.

7.5   Influences of Moving from the Commons to Uncommon Land Tenure

To the nineteenth-century Europeans, the Maasai people were the noblest savages, strangely aristocratic in their disdain of other people's civilization and, seemingly proud and indifferent to all but the most necessary external influences (Amin et al. 1987). They roamed the entire country and interacted with other communities for the sake of their livelihood needs. Droughts made them entice the agriculturalists for food and pastures. For long, nothing proved an inducement to change. Now that the old order is no longer tenable in the new larger nation dispensation that demands sedentalization, the definition of the Maasai pastoralists has changed. The change has made the people retreat to the corner of individualization, consequently being made to use whatever land left to them.

In changing Maasai rangeland tenure, the community complains that it was not appropriately consulted. There is indication that if done, a suitable consensus could have been drawn based on their knowledge of the local environment. The community believes that involvement of others fosters vision and strengthens ties. It is because of this that they respect experts who listen to them as it helps individuals become more valuable parts of the society. The act and practice are informed by the community's proverb *'Meishaa elukunya nabo eng'eno'* that loosely translated infers that one head cannot contain all knowledge (Sankan 1971). As such, stakeholder engagement is reputed to help the community proactively consider the needs and desires of anyone who has a stake in society as it fosters buy-ins for the community's initiatives. The community recognizes that when natural resource users share environmental knowledge, social capital is generated which often results in sustainable resources access and management practices. A shared understanding is found essential to build a cohesive vision, ultimately bringing value to strategic planning and implementing active consultations and engagements. This way, indigenous knowledge is brought to the fore for assessment and eventual embracement in resources development and management.

The colonial government policy to encourage European settlement in Kenya was driven by the belief that a system of interpenetration between the new European settlers and the Africans would be mutually beneficial. A treaty was negotiated with the Maasai to vacate the lands that were required by the settlers and occupy two separate reserves, one to the north of the constructed railway line and the other to the south of it. The Maasai were consequently forced to move to the southern Maasai reserve and away from the most productive areas of their land to make way for white settlers (Hughes 2005). The movement was followed by creation of national parks and other protected areas, further restricting access to critical pastoral resources that altered traditional patterns of land use.

In the 1960s, it became clear that there was insufficient land to allocate individual ranches to all Maasai pastoralists (Graham 1988). The national government subsequently established group ranches by assigning property rights to groups of individuals under committees elected from the membership (Grandin 1986; Munei and Galaty 1998). The joint property rights intended to encourage producers to sedentarize, commercialize, conserve the range, and invest in infrastructure. In turning the land into privately owned parcels through subdivision of the communal land, the government argued that indigenous systems of management resulted in grave land degradation because they encouraged pastoralists to keep excessively large numbers of cattle that outdid the environment's carrying capacity. In the land subdivision many group ranches failed to include year-round grazing. As a result, most families were only able to maintain their livestock herds by moving beyond group ranch boundaries (Kimani and Pickard 1998).

Since gaining independence in Kenya, development among the Maasai in Kajiado has raised acceptance that the ultimate future is that of permanent settlement in definite areas. The development has largely shoved aside the use of indigenous knowledge in access, use and management of the rangeland resources. The change has opened up pastoral lands for Kenya's growing population through privatization where parcels can further be subdivided and sold. The increased land privatization has instrumented erosion of natural resources collaborative management practices which have traditionally been known to increase livestock production and resilience to drought, resulting in socio-economic and ecological marginalization (Western 2009).

The principal land use activity of the Maasai is pastoralism where wildlife grazing alongside livestock enriches pasture composition, mode considered most sustainable for the environment (Tarayia 2004). The aspect favours communal usage of land. The traditional collective land use provided a range of resources used in extensive forms of exploitation. Though land privatization painted a rosy image of land owner rights being secured through private title, the moment of enclosure and privatization has been more often characterized by exclusivity rather than equity. Having a land title deed is reputed to allow one to lease or sell land at will, acquire a loan, have permanent settlement and individual resources management. The new settlers are seen to dominate the pastoral community since they are technologically advanced. The non-Maasai land buyers have little connection to the community, resulting in their ideas and traditions conflicting with the Maasai traditions. Many of the new landowners prefer cultivation that usually leads to conflicts over trespassing. The new landowners do not only share their lands but also put up fences, effectively blocking access to their lands. Solving conflicts in the community the cultural way has become difficult in the mix of foreigners and the local people.

The aftermaths of moving from communal to individual land ownership have altered decision-making processes, social networks, and cooperation of the pastoralists, offering limited access to range resources (Sundstrom et al. 2011). The lessened land sizes limit people from practicing indigenous knowledge based land use and management practices. The problem is heightened by the pastoralists' lack of skills in the management of individual ranches, affecting generation of income, thus reducing availability of food. Drought, rangeland degradation, reduced access to and control of land and an unfavorable political and economic environment are other aggravators of the problems. The reducing number of livestock is seen to inevitably bring about the introduction of other land uses, especially crop production to improve sustainable food security. From the point of view of resource maintenance, ecological sustainability, economic productivity, or community well-being, there seems to be little sound justification for subdividing communal and group holdings in arid and semi-arid lands (Monbiot, 1994). The productivity of rangelands is directly associated with producers achieving economies of scale, making possible the opportunistic movement of herds across scantily distributed grazing and water, and achieving sound management of dryland resources.

7.6   Reinventing the Commons

Within the Maasai community, the majority of the people feel that life under the commons was the best in light of the prevailing resources scarcity. In contrast, there are those that believe that the essences of communal arrangement are irrelevant in the modern globally connected world. Both views have been shaped by real and perceived changes in the society structures. Land subdivision has presented a scare to the indigenous organization of the Maasai livestock herders. The land is further threatened by the effects of climate change, population explosion, increased seasonal failures, disease outbreaks and upsurge in invasive weeds. To this end, some of the community members have opted for alternative livelihoods to sustain the increasingly dire situation, embracing realignments and social, cultural and economic adjustments to enhance their coping ability.

In the Maasai culture, the family often makes decisions on matters of resources that determine livelihoods access, development and manage-

ment. In the model, culture and the extended family play a primary role in matters dealing with survival, and all aspects of life. The indigenous knowledge in the family then becomes a critical asset for posterity. In the 2021/22 drought occasioned by four consecutive years of rainfall absence, the Maasai pastoralists' migration pattern saw them move to the west of their setting as opposed to the traditional eastwards movement. Indigenous knowledge tells that a drought that makes people move to the west would be a severe one characterized by loss of most livestock. The pastoralists who solely subscribed to modern knowledge had a problem in embracing this school of thought and practice, consequently losing massive numbers of livestock.

The Maasai pastoral community has for eons relied on adaptable social networks based on adherence to traditional decision-making systems and reciprocal relations between individuals and communities to help maintain traditional pastoralism. The sedentarization growth reduces the number of households living together in common homesteads (enkang) and increases individually controlled pastures, leading to isolation of individual households and conflict over access to pasture and water resources (Talle 1988). Pretty and Ward (2001) associates the social capital to relations of trust, reciprocity and exchanges, common rules, norms and sanctions, connectedness, networks, and groups. The social networks include bonding ties among families and friends within the community and bridging ties to outsiders for ancillary help and information to the benefit of individual and collective well-being. In the traditional regulated access to common resources like water and grazing land, it occurs in a variety of spatial and social contexts, depending on the resource, group membership, and rights determined by birth, kinships, labor investment, or social contract (Homewood et al. 2009). Though land subdivision has slowed the known and practiced social capital, often, natural resource management strategies are heavily governed by social relationships and networks among individuals and groups.

Faced with land privatization and intensified fragmentation problems, the Maasai pastoralists are resorting to reestablishment of social capital and mutual assistance networks to help maintain access to resources. The formed new networks and alliances reflect the importance of reciprocal use of natural resources (Sundstrom et al. 2011) in a fragile environment in order to maintain access to natural resources necessary for their survival as pastoralists. In many cases, the new mechanisms are based on bonding ties and a shared sense of traditional norms and values. New bridging ties based on access to distant pastures, development opportunities, markets, jobs and educational background are also being developed.

The Maasai households' interaction with others is believed to give direction to processes of collective action and social participation and is a source of common identity. Wilkinson (1991) accentuates a community to be much more than a geographic location as it is a social and psychological entity that represents a place, its people, and the relationships that exist. Recreating the commons for the Kajiado Maasai pastoralists is bringing forth a regrouping dispensation to share resources. This is an indication that while it is true that the Maasai society has changed greatly over the century, the community ideal has not become irrelevant. This is a result of the sense of community that is fostered through interaction, communication, and understanding of common needs that can be met through use of what the people know. Community action is seen as being the foundation of the community-development process because it encompasses deliberate and positive efforts designed to meet the general needs of all local residents. The process represents multiple and diverse interests in the locality, and consequently provides a more comprehensive approach to community development (Wilkinson 1991) intended to benefit the entire community and to cut across diversity that may exist. Corresponding development can be seen as the process of building relationships that increase the adaptive capacity of local people within a common territory to manage, utilize, and enhance resources available to them in addressing their local issues.

7.7 Milestones of Development Failures

Financial, material, time, and human investments have been made in varied projects to enhance the living situation of the Maasai pastoralist people. Yet some of the projects have generally failed to bring sustainable benefits. A critical aspect which has always led to the failure is the lack of adequate stakeholder involvement which resulted in inadequate ownership and commitment. Development support agencies often did not fully consider understanding the socio-economic, cultural, and political factors influencing the projects' design, planning and implementation. Additionally, very limited follow-up support during implementation was tendered by the development agencies. Inclusive and viable community driven approaches to project sustainability could have been achieved through participation and involvement of all stakeholders. Community members report that they needed to have been consulted to identify their own needs, analyzed the factors that led to the needs, and drawn up their action plans to address them. Moreover, respect for and the use of community's intrinsic knowledge and capacities should have been recognized to allow the community to cultivate innovative approaches to address their own problems. At the conclusion of the development supported intervention and the handing over of the projects done by the implementing agencies to the community, the support agencies needed to design and actualize responsible sustainability exit plans to promote continuity of the project. This section shares on six prominent interventions that illustrate failure occasioned by inadequate engagement of beneficiaries to seek their knowledge, skills and opinions.

1. **Group Ranching**

Land tenure changes in Kajiado County have been mostly externally driven, first by the colonial government, followed by the World Bank and finally by the independence government. The three institutions undermined the value of traditional natural resource management that recognize seasonal variations and the need to respect dispersal areas for use in times of drought. Group ranches were formed under the Land (Group Representative) Act of 1968 that provided for the incorporation of representatives of groups recorded as owners of land under the Land Adjudication Act and for the purpose of collective pastoral management and resource use. The concept of group ranches was initially popular among the Maasai pastoralists as it was construed to provide security and safeguard against land alienation by non-Maasai people, and annexation as national parks or government forests (Tarayia 2004). The failure of the system to deliver the objectives of improved livelihoods and security of tenure led to their dissolution and subsequent subdivision. The design of the land sub-division failed to take into account the existing cultural and ecological constraints. At the same time, the government's objectives and targets ignored the objectives of economic efficiency, environmental integrity and equity issues which are core in the pastoralists indigenous land use practices. Consultations with the local communities would have led to embracement of land uses that rhyme with the local pastoralism of appropriate productive land use in line with climate change, reduced land degradation, more equitable access to resources especially water and less conflicts between people.

2. Improved Maasai House Project

The Improved Maasai House Project (IMHP) was initiated on the premise of health concerns and oriented to assist the Maasai pastoralists to have houses that have less smoke, appropriate ventilation and provide appropriate space for school going children to study in. The brought in design embraced high roofs of iron sheets, cement mortar walling and large windows. This was to replace the low roofed traditional clay and dung walls and roofs that were supported by twigs. The indigenous houses were low and allowed the house to be cool during the hot days and warm at nights. The small windows provided a sense of privacy. The improved house building

project intervention failed, the major cause of failure being the lack of recognition of the social, cultural and environmental needs of the Maasai houses. A major source of failure had to do with architecture. Assessments point to existence of acute planning and implementation gaps in the entire process of having the improved houses, the major one being the lack of apposite consultations with the Maasai people. Failure of architects to involve the inhabitants right from the beginning, left owners at stake to accept designs which they never had a complete knowledge about. Consequent inadequate planning and poor implementation strategies groomed serious problems of sustainability. In the end, the improved house would stand empty while the people spent days and nights in the traditional houses.

3. Boreholes Water Supplies

The greater part of Kajiado County relies on groundwater. However, sustainability of the developed water facilities has been a major challenge in many settings. The government and other development agencies have promoted community boreholes without adequately weighing in the sustainability elements. The importance of assessing sustainability in a holistic can enhance the water source wellbeing factors (Spaling et al. 2014; da Silva et al. 2012). Field surveys show that at a given time, less than 30% of the boreholes are not functional, the source of failure being social, technical, financial, environmental and institutional factors (Mwangi 1991). Some of the boreholes have failed due to inadequate geological investigations that would have benefited from the local knowledge on groundwater occurrence and the history of deep groundwater development. Some of the introduced pumping equipment are not able to operate in the high temperature and dusty environments that are characteristic of the ASALs. On the other hand, the boreholes user groups were not provided with appropriate and adequate managerial skills to sustain the water supplies. Feedback from the community shows that appropriate consultations with the Maasai pastoralists would have led to a choice in improvement of the Maasai traditional shallow wells *(Ilumbwa)* which are developed along the seasonal rivers (Mwangi 1991; Rutten 2004). The people have for eons been with the shallow groundwater sources technology and understand it best, and the operation and maintenance costs are affordable.

4. Environment Protection Interventions

In the 1980s and 1990s, Kajiado County received government and other development support agencies interventions towards environmental conservation projects. Most of the activities were directed towards tree planting and perennial water catchments protection. External knowledge was used to effect the desired outcomes that were largely programme based and not community. This was a time when land sub division was also taking place in the group ranches and people were becoming more individualistic and the vulnerability of the Maasai pastoralist getting more exacerbated by the interaction among multiple stresses that included poverty, land use change resulting in low adaptive capacities (Maito and Odhiambo 2013). Planned adaptation actions were therefore needed to respond to existing and anticipated impacts of climate change and variability. Communal conservation measures were rapidly disappearing as the areas of collective interest were either falling to government or individual interests. As such, indigenous knowledge of the people on environmental matters and the linkages was being increasingly sidelined. The planted trees which were largely exotic could not survive and the water catchments became encroached because of personalized and political interests. The extension staff that was promoting the environment conservation ideals had major difficulties in convincing the pastoralist to absorb what was considered ideal to build the environment space in an ASAL circumstance where fluctuations and variations in climate, particularly rainfall and temperature, adversely affect the physical, biological and socio-economic systems leading to disasters and calamities (Tasokwa 2011). To the Maasai, the environment interventions should have concentrated on developing more water sources, suggesting a need for community endowment with a wider range of information and capacities upon which they could rely on to make choices on environmental concerns.

5. Ferroement Water Jar Project

The initiative applied the chicken wire and cement mortar technology to harvest rainwater from the traditional Maasai house roofs and other erectable surfaces. The project was promoted by a foreign agency as a public health endeavor as part of reducing the walking distance to watering points. Many households were involved without much success as the quantities collected were not adequate to meet the domestic needs that included watering livestock. The other challenges were inadequacy in the housing architecture and the erratic rains. The heaviest problem however was the migration pattern of the pastoralists who would move away during drought to track rains and pasture. At times the houses would be demolished and the water jars left standing in the wilderness with no water collection appurtenances. The Maasai pastoralists consider lack of water *(enkare)* primarily as a problem for their herds and less as a matter of concern regarding their own personal consumption (Rutten 2004). The extension staff in the project did not verse itself on the water needs and utilization patterns of the people and the seasonal migration demands. Neither did they involve the people, particularly the women who are culturally vested with house constructions in designing the water storage facilities. The women make leak-proof mud and dung walls and roofs and would probably have inputted their knowledge in the water jars inventiveness. Plastic water storage tanks that were later introduced in the area as an alternative and became a major success as the people would migrate with the water facilities and the water quantities were rationalized with household needs.

6. Introduction of Camels among the Maasai Pastoralists

The Maasai livestock herd is traditionally composed of cattle, goats and sheep. The animals are normally hard hit by droughts. Camels were introduced to the Maasai community in 1989 by a European development support organization in liaison with the Kenyan Ministry of Livestock Development as part of diversifying livestock in efforts to raise resilience building against climate change shocks. Food security was seen as an important advantage in having the animal in the community. The camels which were brought from northern Kenya were in the end taken by the Maasai pastoralists as a foreign animals that did not fit within the local resources management demands. The animals were accused of being dirty as they urinated on themselves, would drink water from the water pans and decide to lie in there, destroyed the dry season grazing portions near the homesteads *(Olopololi)* and ate whole plants as opposed to the Maasai livestock. The introduction targeted the Maasai women who culturally never owned land. Men were sidelined in the project identification, which was a mistake for a patriarchal society. As such the women were to seek permission from men to keep the animals and access water and pasture resources. The men saw the project as belonging to the women and thus gave very little support, compromising the sustainability of the animals' upkeep in the community. Added to this, the lack of experience on camel husbandry and animal diseases were major challenges faced by the Maasai pastoralists, the most pronounced being the camel trypanosomosis disease (Bukachi et al. 2003). Other important constraints affecting camel production included lack of markets for camel products, lack of pasture, inadequate veterinary services and inadequate farmer training involving camel handling skills, disease identification and management. Eventually the project collapsed and all the camels were sold away. The accrued finances were directed by women to steer fattening which the local people know how to effect.

7.8   Pursuing the Missed Path to the Promised Land

This chapter reveals some of the problems emanating from planning for the Maasai pastoralism setting in a national governance system committed to disregarding people's knowledge and embracing what is externally concluded as the panacea for the rangelands. The consequent development actions have worked to render indigenous pastoralism unviable, swelling the risks to continue with pastoralism. The land tenure interferences have upset community solidity and generated disagreements over access to range resources between adjacent landowners and the communities (Sundstrom et al. 2011). The incoming Maasai rangelands land procures bring in backgrounds and new land manipulations that are at variance with the local pastoralists' way of life. Many Maasai pastoralists' landowners are embracing the new dispensations and, ultimately becoming increasingly individualistic. To some of the pastoralists, the land tenure and use vicissitudes represent auspicious livelihoods diversification and hence a better future in light of the reduced access quality and quantity to range resources. Others find the changes as an interruption of equivalence to evictions and forfeiture of livelihoods, making life unbreakable by ignoring their reality and the knowledge that has seen them survive in the milieu.

To realize the global Sustainable Development Goals (SDG) agenda, policies and education systems must encourage greater respect and cohesion among people and their relationship with nature. In these efforts, indigenous knowledge should be paid greater consideration and embraced as part of the principles of sustainable living. Confronting the challenges faced by the pastoralists requires policies and interventions that not only address the direct challenges they face but also reverses the negative prejudices that often portray pastoralism as primitive, unproductive and environmentally destructive. One of the concrete ways of responding to the challenges and myths about pastoralism is through effective information and knowledge exchange that the plug-in concept posits, reiterating the exchange must start from a good understanding and appreciation of the indigenous knowledge of the people. Conventional development processes must build on the prevailing indigenous knowledge to ensure usefulness, success and sustainability. The use of knowledge platforms with indigenous knowledge at the centre with tailored and targeted messaging can increase the use of evidence-based policy and program decision making to promote dialogue among various actors relevant to pastoralism aimed at raising awareness on the important contributions of pastoralists to social, economic and environmental wellbeing of the country. Considering the regional and localized wisdom of indigenous knowledge groups, the capacity of the subject communities to contribute to sustainable development, governance requires a bottom-up approach in which local institutions and community organizations are empowered in higher-level dialogues. Integrating indigenous knowledge and other relevant knowledge systems in development and governance models will empower ignored sustainability experts in shaping a shared future in several ways. Programs that communities are willing to accept and embrace are likely to depend on indigenous knowledge factors as problems and potential solutions are likely to be defined in a manner consistent with the local philosophies.

Using indigenous knowledge concepts and principles in development among the Maasai pastoralists and then plugging in other development principles is strategic to ensure strong political commitment to be able to achieve sustainable and broad-based changes and improvements. Appreciation of this has potential to break down prejudices against the knowledge and fusing it in development principles and, advance applicability to problems of finding mutual understanding, common ground, and consensus of sustainably managing and sharing resources.

References

Amin M, Willetts D, Eames J (1987) The last of the Maasai. C. Struik

Arid and Semi-Arid Lands Programme (1988) Kajiado Boreholes Survey, ASAL Programme, Kajiado

Breman H, de Wit CT (1983) Rangeland productivity and exploitation in the Sahel. Science 221:1341–1347

Brennan MA, Kumaran M, Spranger M, Cantrell R (2013) The importance of local community action in shaping development. Institute of Food and Agricultural Sciences, University of Florida

Bukachi SA, Chemuliti JK, Njiru ZK (2003) Constraints experienced in the introduction of camels in Tsetse Fly infested areas: the case of Kajiado District, Kenya. J Camel Pract Res 10:145

da Silva FE, Heikkila T, de Assis F, de Souza F, da Silva DC (2012) Developing sustainable and replicable water supply systems in rural communities in Brazil. Int J Water Resour Dev 29(4):622e635

Gadgil M, Berkes F, Folke C (1993) Indigenous knowledge for biodiversity conservation. Ambio 22:151–156

Graham O (1988) Enclosure of the East African Rangelands: recent trends and their impact, pastoral development network paper 25a. Overseas Development Institute, London

Grandin BB (1986) Land tenure, subdivision and, residential change on a Maasai Group Ranch. Bullet Inst Dev Anthropol 4(2):9–13

Homewood K, Kristjanson P, Trench PC (2009) Changing land use, livelihoods, and wildlife conservation in Maasailand. In: Homewood K, Kristjanson P, Trench PC (eds) Staying Maasai? Livelihoods, conservation and development in East Africa Rangelands. Springer, New York, pp 335–368

Hughes L (2005) Malice in Maasailand: the historical roots of current political struggles. Afr Aff 104(415):207–224

Kariuki PM, Onyango CM, Lukhoba CW, Njoka TJ (2018) The role of indigenous knowledge on use and conservation of wild medicinal food plants in Loita Sub-county, Narok County. Asian J Agric Exten Econ Sociol 28(2):1

Kereto J, Nkurumwa AO, Obara J, Mango N (2022) Livestock management and protection using indigenous technical knowledge among the Maasai of Narok County, Kenya. Cogent Soc Sci 8:204079

Kimani K, Pickard J (1998) Recent trends and implications of group ranch sub-division and fragmentation in Kajiado District, Kenya. Geogr J 164(4):202–213

Kotut L, McCrickard DS (2021) Trail as heritage: safeguarding location-specific and transient indigenous knowledge. In: 3rd African human-computer interaction conference (AfriCHI2021), March 8–12, 2021, Maputo, Mozambique. ACM, New York

Maito TL, Odhiambo EOS (2013) Pastoralism: a livelihood system in conflict. Asian J Manag Sci Educ 2(4):249

Mihessuah DA, Wilson AC (Eds) (2004) Indigenouzing the Academy: Transforming Scholarships and Empowering Communities, Lincoln

Monbiot G (1994) The tragedy of En- closure. Sci Am 270(1):159

Munei KO, Galaty JG (1998) Maasai land, law, and dispossession. Cult Surv Quart 22(4):68–76

Mwangi M (1991) Alternatives to boreholes: some thoughts, paper presented at the workshop on the Kajiado District Boreholes Survey, Namanga, Kenya, March 25–26. ASAL Programme, Kajiado, Kenya

Ndago P (2024) Land use diversification in Elangatawuas area of Kajiado County, unpolished project report. South Eastern Kenya University

Pretty J, Ward H (2001) Social capital and the environment. World Dev 29(2):209–227

Richards P (1985) Indigenous agricultural revolution. Hutchinson, London, UK

Rutten M (2004) Shallow wells: a sustainable and inexpensive alternative to boreholes in Kenya, Paper presented at the EU conference 'Support to Marginal Rural Areas in Somalia', Nairobi, 23–26 November 2004

Sankan SS (1971) The Maasai. Kenya Literature Bureau, Nairobi

Skinner D (2010) Rangeland Management for Improved Pastoralist Livelihoods: the Borana of Southern Ethiopia. Oxford Brookes University

Sow S, Ranjan S (2021) Indigenous Technical Knowledge (ITK) for sustainable agriculture in India. Agric Food: E-Newslet 3(1):31. Researchgate.net/publications

Spaling H, Brouwer G, Njoka J (2014) Factors affecting the sustainability of a community water supply project in Kenya. Dev Pract 24(7):797e811

Sundstrom S, Tynon JF, Western D (2011) Rangeland privatization and the Maasai experience: social capital and the implications for traditional resource management in Southern Kenya. Soc Nat Resour 25(5):483–498

Talle A (1988) Women at a loss: changes in Maasai pastoralism and their effects on gender relations, Stockholm studies in anthropology No. 19. University of Stockholm, Stockholm, Sweden

Tarayia GN (2004) The legal perspectives of the Maasai culture, customs, and traditions. Ariz J Int Compar Law 21:184–222

Tasokwa VMK (2011) The impact of climate variability and extreme weather events on gender and household vulnerability to food insecurity., PhD thesis,. University of Nairobi

Warren DM, Rajasekaran B (1993) Putting local knowledge to good use. Int Agric Dev 13:8–10

Warren, D.M., Slikkerveer, J. Brokensha, D. (1995) 'Introduction' in the cultural dimensions of development: indigenous knowledge systems (eds. M.D. Warren, LJ. Slikkerveer W. Brokensha, Intermediate Technology Publication, London 15–18

Western D (2009) The drought of 2009. Amboseli Conservation Program. http://www.amboseliconservation.org/amboseli-drought-2009/

Wilkinson KP (1991) The Community in Rural America. Greenwood Press, New York

World Bank (1999) What is indigenous knowledge? www.worldbank.org/afr/ik/basic.htm

Evaporating Indigenous Knowledge Based Justice System Among the Maasai Pastoralists of Kenya

8

Ida Gathoni

Abstract

Kenya is a heterogeneous society that comprises over forty ethnic groups. The diversity validates the need for recognition of rationalized individual community's justice dispensation arrangements by the country's judicial system. This Chapter shares on how changes in the country's governance has affected acknowledgement of participation of the informal justice indulgences to assure equity. The conversation is based on experiences among the Maasai pastoralists community of Kajiado County that inhabit the South of Kenya to explore the implications of the presence or absence of appreciation and housing of the peoples' justice system and the resultant consequences. The study is also informed by literature review, official reports and structured questionnaires administered to key opinion leaders and selected household heads. A conclusion is that while the indigenous justice system is remedial and restorative of wrong-doers into the community, the formal one is bent on punishments which place corporeal and psychological strains on reprobates. Illuminated is that although the indigenous ways of bestowing justice is marginalized by the conventional one, its approval has received heightened attractiveness among the Maasai pastoral community as an alternative legal pathway, principally driven by the distress arising out of the complications experienced in the government courts. Outcomes present the indigenous justice system as viable and relevant for the pastoral community, justifying its place in the national legal framework. The study vouches for supporting the indigenous legal institutions their justice mandates, building their capacity and, linking them to the formal justice system can improve access to reasonableness in arriving at palpable justice.

Keywords

Indigenous · Customary · Informal · Statutory · Restorative · Alternative

8.1 Introduction

The Maasai pastoralist community have traditionally thrived on periodic relocations in search of pasture and water for their livestock. The practice has been demised by the government ultimatum to sedentalise, a move that has among others subjected the Maasai people to inadequate range resources and a need to wholly subscribe to the national justice dispensations which the people regard as insufficient in providing 'fairness'. The

I. Gathoni (✉)
Kenyatta University, Nairobi, Kenya

© The Author(s) 2025
S. Dittoh et al. (eds.), *Integrating Indigenous and Scientific Knowledge for Sustainable Food Systems in Africa*, Sustainable Development Goals Series,
https://doi.org/10.1007/978-3-031-85512-2_8

problem is aggravated by the country's ill staffed and equipped courts for apt legal deliverables.

Traditionally, the Maasai people have had ways to justifiably effect justice. Vicissitudes orchestrated by alterations in the country governance system have however subordinated the community's justice system and oriented them to the state legal constituency. Getting edified on how the community currently realizes justice and the place of their indigenous legal institutional set up and knowledge in the space has potential to inform and lobby policy. The implications of conflicts that arise between customary and formal justice systems and how they can be overcome becomes a contributor in the debate on having a legal framework that can accommodate multiple systems to effectively serve the community injustices resolution.

Reforms and development calls around the globe are on the rise, particularly on non-negotiable issues of the rule of law. Campaigns exist about legal heterogeneity based on different world views and sources of validity, maintained by forms of organization other than the state (Benda-Beckmann 2001; Adoko and Levine (2009); Fayo 2011).

While rooting for limited judicial enforcement of constitutional guarantees for customary laws, Pimentel (2010) argues that legal pluralism regime, customary courts and customary law can become guardians not only of traditional culture, but also of rule of law principles and human rights. The indigenous knowledge based justice systems find strength in their derivation, being home grown and thus found to be culturally appropriate by the communities they serve.

Oomen (2005) opines that the law of the state needs to reflect the diversity of culture and alternative existence of cultural laws. The recognition is foreseen to offer an avenue to front different legal systems that can mutually exist (Pimentel 2010). The rationale is supported in Zehr (1990) who reiterates on stakeholder participation in supporting the justice process at the national and customary arrangements. Watson (2001) and Fayo (2011) orates on cognizance that institutions are not static and hence the need to provide ability to act upon changes. As such, the state and

its subjects must recognize each other's desires in the justice provision spectrum.

8.2 The Maasai Indigenous Justice System

Leila et al. (2005) enlightens that human behavior is shaped and influenced by living normative frameworks inherent in customary laws which are indigenous in an environment and ingrained in the customs and traditions of people. The aspect is experienced among the Maasai community who despite the dominance of the formal legal system in Kenya, many of the community members resort to the indigenous knowledge based justice arrangement because of the difficulties associated with the formal justice system. Legalese, high cost, inaccessibility and lengthy legal procedures are cited as discouraging points in approaching the formal justice system. In contrast, the customary based justice dispensation is reputed to be locally brewed and thus understandable, embraceable, accessible and within local economic means. The community gives the indigenous legal avenue credence as being foremost restorative, expeditious and supple, aspects that are absent in the formal justice system. Restorative justice ensures that the burden of shame, guilt and pain is shared among the community and the victim (Fayo 2011).

The Maasai indigenous knowledge system carries with it rules and regulations that particularly define access and rights (Sankan 1971). Before land individualization, it was communally owned, taken care of and accessed. Local culture dictated collective responsibility on the land and the therein resources and therefore all were obligated to protect it as a common good. Specific bushes were reserved as sources of firewood, water catchments, ceremonies, and herbs. The settlement pattern was agreed on to avoid inconveniences. Informed by the wisdom of elders, the community indigenous institutions guided in using and managing the common resources as well as dispensing justice where necessary. The customary rules governing the right of land tenure were designed and sanctified for the community's sustainable survival and placed no monetary value on land.

8.3   Administration of Justice Among the Maasai People

The Maasai pastoralists' justice standards are premised on the principle that members of the community are versed with the requirements of the community for fairness to consequently desist from disrupting justice and seeking appropriate redress. Existence of communal participation in justice dispensation allows injustices to be viewed as a blunder that requires corrective intervention, resulting in offenders remaining integral to the community (Sankan 1971; Fayo 2011)). The resultant restorative justice views injustice as a transgression against people. The informal system therefore creates an obligation to make things right (Zehr 1990). Restorative justice finds its base in the culture and tradition of the people and is thus related to how people think about the law-breaking which constitutes to how response is provided. Restorative justice has been a way of life for indigenous societies in resolving conflicts and delinquencies to maintain cohesion and accord in the societies. The participatory approach applied in providing justice succeeds in arriving at mutual agreement between victims and wrongdoers through the lenses of having a greater good of the society (Fayo 2011).

The Maasai pastoralists' foundation for their justice system is their culture, which the whole community is obliged to maintain and sustain. The cultural values prescribe repairs to damage to persons, offering social support and security. The justice institutions are linked through clan and moiety relationship (Sankan 1971) with most cases being dispensed off at the clan level. The justice sessions allow community contribution towards the problem identification and solution provision. The justice dispensation is guided by recognition that all community members are equal before the law irrespective of position and status. As such, the entire community is subjected to the same punishment for violating the laws.

The Maasai pastoralists' customary justice underlying principle is similar to the one practiced among the Borana pastoralists (Fayo 2011) which underlines reparations and restorations. The two core aspects of the judicial arrangement are reputed to inculcate trust in the indigenous justice institutions. The fronted reasons are that in case of injustices, the committer and the victim have a right to benefit from the social support that the community structure accords its members. The defense and the claimant have to be present for a case to proceed and once a verdict of guilt is arrived at, the perpetrator is urged to orally admit as a mien of repentance. Repeated offences can lead to banishment from the community, a consequence that can affect the offender lineage. It is the responsibility of every community member to ensure that at all times the offender does not get back to the crime again (Fayo 2011).

Various livelihood fines are meted for wrong doing. Traditionally, theft of a bull is allocated a fine of five cows while that of a heifer is seven. In some Maasai moieties, stealing a sheep invites a fine of four sheep while the majority charge two. In the event of a marriage breaking down, the cattle bride's worth is repaid in original numbers, with a substitute of two sheep being regarded as equal to a heifer.

When a man accidentally kill another's donkey, the fine is decided by elders. Normally a lamb and a heifer payment applies. An individual apprehended stealing from a hive is placed under a fine of two sheep or a heifer. Notably, until recently, cultivation related fines did not exist among the Maasai as traditionally the community was not into land tilling. Related forms of fine have surfaced with a charge of two sheep or one heifer being meted on people who steal from someone's field. In case an adolescence is involved, the fine is one sheep. Councils of elders have been set up in some crop growing areas to adjudicate cases. One common set rule is for cultivated fields to be protected against livestock intrusion. If somebody fails to erect a fence and livestock strays into the field, the fault is regarded as that of the land owner and consequently no compensation is employed. The council of elders allows fines based on the volume of crops damaged by livestock in fenced off fields. The Maasai pastoralists usually set aside pasture conservation portions *(Olopololi)* near homesteads. Trespassing is not allowed as a measure to ensure presence of pasture in stressful times. When a livestock gets into someone's *olopololi,* negotiations customarily take place between the people affected and an amicable position arrived at. If someone steals another's water, a fine is imposed, which depends on what elders agree on.

The conception of the guilt of murder among the Maasai does not extend beyond the borders of Maasailand. Thus a man can only be regarded as guilty if he murders another Maasai (Sankan 1971). When one murderers a man, the fine *(enkiro)* is forty-nine cattle. The forty cattle are negotiable, but the nine must be included, the nine being the number of orifices in a man's body. Traditionally, there are no fixed fines for the murder of women. However, a common fine which is only applied in killing of a Maasai woman is comprised of forty eight or twenty eight sheep that are given to the father of the woman or her relatives. The figure 'eight' is related to the things normally associated with women, a loin-cloth, gut for repair work, a needle, a calabash, a razor, an axe, a reed for cleaning calabashes and cowrie-shells. If a man kills a woman by mistake, he must undergo a ceremony of expiation in which he is cleansed and purified because it is thought that the dead would bring a curse upon the man unless this is done. If an uncircumcised boy is murdered no fine is imposed until the age group is circumcised. The age fine is forty-nine sheep. A fine of one heifer *(ilabaar, enkipito)* is imposed on anyone who causes someone else to have a fracture. This fine has to be paid quickly because if the injured man dies even from another cause before the fine is paid, the offender will be sued for murder.

Inheritance is part of the social justice arrangement in the Maasai community. The recent state law of inheritance that considers all family members as equal in property rights is yet to take place in the Maasai society, though a portion of the population has been seen to consider it. Livestock is still the actual property to be inherited and shared at the household. Tarayia (2004) enumerates the basic items of inheritance at the family level to be the father's traditional stool *(olorika loo nkejek)*, snuff and tobacco container *(olkidong)*, metal bracelets *(ilkataarri)*, sword *(olalem)*, ear ornaments *(isakankarri anaa muna)*, walking stick *(olartat)*, and cloth made from columbus monkey or hyrax skin *(enkila)* that is only worn at occasions. She reiterates that the girl-child is considered to settle down with an own husband, children, and livestock and therefore, not qualify for her parents' property. However, the father has the authority and the right to apportion anything to her in his estate. With the changes in land tenure, land can now be inherited revealing that with the onset of land privatization, land laws no longer conform to customary family practices (Tarayia 2004).

Judicial matters in the traditional sense are the preserve of elders (Tarayia 2004). The making of a Maasai justice dispenser follows the community's democratic procedures to engage individuals to the stool of justice provision, whose graduation is partly determined by going through prescribed ceremonies that are an expression of the people's culture and self-determination. The common denominator in the ceremonies is provision of fairness, respect and responsibility. The key ceremonies that one has to undergo include circumcision *(emuratta)*, marriage *(enkiama)*, warrior-shaving *(eunoto)*, and, junior elder *(olngesherr)*. Three core leaders—*Olaiguanani lenkashe, Oloboru enkeene* and *Olotuno* - are chosen at the *eunoto* stage to shoulder age set's bad and good deeds. The leaders are bound by the collective necessity to dispense justice to all and follow all laid out aspirations of the entire community wellbeing (Sankan 1971).

8.4   Contrasting the Customary and Formal Statutory Justice Systems

Reflections on the Kenyan legal arena identifies incompatible characteristics between the customary and formal justice standings. This is despite the country constitution directing on having affirmative action to appreciate cultural values

(National Council for Law 2010). Despite this, the country is more inclined towards the formal legal system, a development that makes the Maasai people construe the state law as imposed and of little to do with the community's way of life. The country is realized to promulgate laws that are not well-matched with people's lives and livelihood practices. The absence of all-encompassing legal frameworks reduces the grassroots understanding of justice and how to execute the same through customary conducts. The country judiciary does not object to the community usage of customary justice systems but reiterates that if the case arrives at the government courts it would be determined as per state law.

Legislation without the input of the local communities is deduced to be the cause of cynicism in the interpretation of the law as peoples' practices are found to be inconsistent with national policies. The Maasai community believes that laws are not to be made by single institutions such as the legislative assembly but by the entire society with full recognition of prevailing cultures. The prescription is projected to avert legal flaws such as the former communal land tenure where the government considered land to be a trust and thus could appropriate it at will. The community legal elucidation was that the land was rightful theirs based on rights rooted in their traditions rather than legal statutes. The other contentious piece of law is the water act which bestows the ownership of the water to the state (Mumma 2005). The bone of contention is why the pastoralists need to get consent from the government to use or develop water sources they consider their own.

There has been failure by colonial and independent governments in Kenya to understand the dynamics of the Maasai pastoralists' people and their pastoralism. Consequent formulation of legislation and policies have not been in tandem with the customary ways. For one, pastoral development policy by the colonial government was influenced by the view that communal rangelands pastoralism encouraged environmental degradation (AU 2010). Consequently, a sedentary life was imposed and the pastoralists stripped of their rights to large portions of their rangelands, which were given to the colonial administration. The reduced access to resources conflicted with the seasonal migration of the people and their livestock.

Among the reasons that make the Maasai people cling to their customary means to claim justice are as indicated in Table 8.1. The biggest percentage (87%) of respondents find absence of restoration as a major legal flaw and absence of humane way to provide justice. The formal courts process is also found outlandish to the people, not easily accessible and the language and ways of the courts are not easily comprehensible. They are also accused of foot dragging in arriving at decisions and being expensive. There is the feeling that the formal courts are subject to corruption and thus justice can be compromised by the

Table 8.1 Justification for dependence on customary justice system

Validation		Responses [N = 30]
1.	Formal courts process is alien	18
2.	Formal courts are not easily accessible,	16
3.	Not understanding the language the formal courts	19
4.	Not understanding the ways of the formal courts	21
5.	Formal courts are time consuming	15
6.	Formal courts are very expensive	17
7.	Formal courts are subject to corruption	23
8.	Going to formal courts undermines the elders institution	16
9.	Formal courts don't help to restore broken relationships	26
10.	Jailing of offenders by the formal courts invites family suffering	20

Source: Gathoni (2021)

highest bidder. Some of the respondents find jailing of offenders to worsen the social and economic position of the family of the jailed person and therefore an unfair way of treating malefactors. The deduced disadvantages have the capacity to make the pastoralists want to cling to their indigenous legal system. Melton (2004) indicates that indigenous communities may not embrace the formal justice paradigms as they are a difficult choice for those with a functional indigenous justice system.

8.5 Plugging into Indigenous Knowledge for Legal Heterogeneity

Having a justice system that embraces legal logic diversity is vouched for as it recognizes coexistence and value of frameworks that accommodate written and unwritten norms laws (Pimentel 2010). While according credit to the legal multiplicity, Ribot (2008) advocates for decentralization of judicial systems to allow having room for alternative justice resolution mechanisms. Oomen (2005) campaigns for legal plurality, emphasizing on state recognition of cultural diversity and the consequent existence of alternative law governance expressed in terms of social and cultural differences. The resultant respect is reputed to particularly benefit the natural resources dependent populations such as the pastoralists in sustainable access and claims.

The Maasai indigenous justice system finds base in the community's social arrangement and traditional concepts to life. Individuals are part of the community and its system and, consequently have duties to adhere and perform within the confines of the local justice dispensations. Every person has equal value and worth. As such, institutions and their responsibilities and ideals must not just be imported from outside but must be understood in local contexts.

Cognizance is notwithstanding the social and economic transformations that have taken place among the Maasai pastoralists of Kajiado in the century, the community cultural beliefs, ideals and philosophies prevail, marked by concomitant procedures on judgments. Endeavors are driven by the need for attainment of fairness in the community that rhyme with Buckley (2006) prompt that laws and justice are part of the fabric of a community and justice not only envelops the individual and society, but also resides within the individual. The Maasai pastoralists believe they are all good people but at times ignorance or bad luck lead them to going against set norms. Subsequently, the applied justice framework demands administration of fair-mindedness in a way that rehabilitation prevails and harmony is sustained. The fairness dispensation process can be compared to the McNaughton rules that designate an offence committed by an individual who at the time of doing it may be as a result of unsound mind and thus incapable of knowing the involvement is either wrong or contrary to the law (Asokan 2007).

The majority of institutions for resolving injustices in Kenya are informal, largely revolving around customary and religious institutions embedded in respective societies. The foundations are however overlooked when the government is averting injustices or reinstating stability. Credence is instead conferred to the formal justice sectors in the belief that informal legal practices are noninclusive and counter to human rights.

Though the Kenyan governance system has been reluctant to actualize accommodation of indigenous knowledge justice schemes, 94% of the respondent Maasai pastoralists find it imperative for the formal justice system to engage their customary justice practices and its decisions framework. The engagement of the two systems is reputed to offer higher promises, both for the preservation of cultural values and institutions, and ultimately for the establishment of the rule of law. Ribot (2008) believes legal decentralization is a good alternative justice provision without substituting the central government judicial systems. The formal and customary justice systems have a consumer commonality that invites both systems to accommodate one another and coexist within a legal framework that accommodates both in realizations of justice (Pimentel 2010). Failure to bridge the relationship gaps is forecasted to undermine the roles of the two legal systems in discharging respective responsibilities. Among the Maasai people of Kajiado, land is a putative issue, as evidenced by the large number of cases in government courts whose resolutions take long. To expedite and reduce the backlog in the land cases, the formal corridors of justice are slowly recognizing the necessity to allow an alternative justice system. Accordingly, services of some elder panels have been sought to receive and determine land disputes. Occasionally, conflicts ensue between the state and customary laws which communities navigate and find workable solutions.

Colossal value is detected in coalescing informal and conventional legal systems. The merged legal knowledge is foretold to afford a sustainable risk minimizing mechanism that identifies related institutions as enablers of resolving justice problems (Warren et al. (1995); Chambers (1997). The quest by the Maasai pastoralists for appreciation of their traditional justice system is part of a self-determination to control their lives within a particular jurisprudence (Sankan 1971). Emanating from the weighty cultural links that exist between the people, their land and allied resources, there is a need for concerted understanding of the relationship. The prayer to the state is a realization that the country has no monopoly of justice and needs to recognize the existence of customary laws and the need to negotiate with them (Fayo 2011).

References

Adoko J, Levine S (2009) How can we turn legal anarchy into harmonious pluralism? Why Integration is the Key to Legal Pluralism in Northern Uganda. In: Paper Presented the United States Institute of Peace, George Washington University and World Bank Conference on Customary Justice and Legal Pluralism in Post-Conflict and Fragile Societies, (17–18 November 2009)

Africa Union (2010) Policy framework for pastoralism in Africa – securing, protecting and improving the lives, livelihoods and rights of pastoralist communities. Ethiopia, Addis Ababa

Asokan TV (2007) McNaughton (1813–1865). Indian J Psychiatry 49(3):223–224

Benda-Beckmann F (2001) Legal pluralism and social justice in economic and political development. IDS Bull 32(1):46–56

Buckley K (2006) The externalization of justice. In: Paper Delivered at the 7th Annual Conference on Evil and Human Wickedness; Salzburg Austria. March 13–17, 2006

Chambers R (1997) Whose reality counts? In: Putting the last first. Intermediate Technology Group, London

Fayo GD (2011) Coping with scarcity in Northern Kenya: the role of pastoralist Borana Gada Indigenous Justice Institutions in Conflicts Prevention and Resolutions for Range Resources Managements. In: A Research Paper. International Institute of Social Studies, The Hague

Gathoni I (2021) Conflicts and cooperation over natural resources. Unpublished Field Report, Kajiado, Kenya

Leila, C., Sage, C and Woolcock, M. (2005) Customary law and policy reforms: engaging with the plurality of justice systems. http://namati.org/resources/customary.law-and-policy-reform-engaging-with-the-plurality-of-justice-system/

Melton AP (2004) 'Indigenous Justice Systems and Tribal Society', National Institute of Justice, USA

Mumma A (2005) Kenya's new water law: an analysis of the implications for the rural poor. In: Van Koppen B, Giordano M, Butterworth J (eds) Community-based water law and water resource management reform in developing countries. CAB International, Wallingford

National Council for Law (2010). Constitution of Kenya, 2010, Nairobi, Kenya

Oomen B (2005) Chiefs in South Africa: law, power and culture in the post – apartheid era. James Currey Ltd, Oxford

Pimentel D (2010) Can Indigenous justice survive? Legal pluralism and rule of law. Harvard International Review, Cambridge

Ribot J (2008) Legal pluralism and decentralization: natural resource management in Mali. World Dev 36(11):2255–2276

Sankan SS (1971) The Maasai. Kenya Literature Bureau, Nairobi

Tarayia GN (2004) The legal perspectives of the maasai culture, customs, and traditions. Ariz J Int Comp Law 21:184–222

Warren DM, Slikkerveer LJ, Brokensha D (eds) (1995) The cultural dimension of development: indigenous knowledge systems. Intermediate Technology Publications, London

Watson E (2001) 'Inter institutional alliances and conflicts in natural resources management': preliminary research findings from Borana, Oromiya Region, Ethiopia. In: Marena research project, *Working Paper* No. 4

Zehr H (1990) Changing lenses: a new focus for crime and justice. Herald, Scottsdale

Challenges to the Use of Traditional Ecological Knowledge in Natural Resource Management in Rural Eastern Cape, South Africa

9

Chenai Murata and Gladman Thondhlana

Abstract

Although it boasts a rich historical wealth of traditional ecological knowledge, the rural Eastern Cape of South Africa continues to be an environmental hotspot since the mid-twentieth century, suffering land degradation, overharvesting of resources and water scarcity. Over the years, much research effort has been directed to investigating how local communities use traditional knowledge to manage resources. By and large, the studies argue that traditional knowledge provided effective management strategies in the past, hence communities should continue using it. This raises the question: what is happening with traditional ecological knowledge in contemporary rural Eastern Cape in South Africa? To address this question, this work used snowball sampling to recruit 83 respondents from three villages of Colana, Gogela and Nozitshena for in-depth interviews. The respondents were elderly women and men whose families had at least a two-generation history of living in these villages. Using the meta-theory of critical real-ism, particularly its analytical tools of abduction, hermeneutics, dynamism and context, this work found that there are multiple factors, exogenous to traditional ecological knowledge, that make it fail to effectively address environmental challenges in the present-day rural Eastern Cape. They include changed worldviews, legal pluralism and changed lifestyles which are a consequence of adverse incorporation of traditional systems into the South African nation-state system. Indigenous practices, traditions, local ethics, taboos, social conventions, and other cultural practices, that ensured the usefulness of TEK, have been abandoned over time. Incorporation of rural traditional systems into the South African nation-state have been largely adverse. There has been little or no integration of TEK and the "imported" lifestyles. The application of the "plug-in" principle in this case, thus, faces a very formidable task despite the reality that it is the way to go to ensure sustainable natural resources management and the achievement of some of the SDGs.

Keywords

Traditional ecological knowledge · Eastern Cape (South Africa) · Critical realism · Concept of *isizwe* (the nation) · Legal pluralism · Plug-in

C. Murata (✉)
University of Vienna, Vienna, Austria

G. Thondhlana
Rhodes University, Makhanda, South Africa

© The Author(s) 2025
S. Dittoh et al. (eds.), *Integrating Indigenous and Scientific Knowledge for Sustainable Food Systems in Africa*, Sustainable Development Goals Series,
https://doi.org/10.1007/978-3-031-85512-2_9

9.1 Introduction

This chapter uses a South African case study to join and hopefully contribute insights to the growing global socioecological debate on why indigenous knowledge system (IKS), especially its environmental branch called traditional ecological knowledge (TEK) was applied in the past and effectively solved environmental challenges but struggles in contemporary communities. One way of joining the debate would be to examine the efficacy, that is strengths and weaknesses, of TEK. The other one is to investigate how TEK is perceived in contemporary communities which should apply it: an approach that this chapter adopts. We argue that a necessary precondition for any knowledge system to be judged efficacious or otherwise is to establish if the knowledge is embraced and fully applied, otherwise such judgements become unfair and inappropriate. Accordingly, the chapter focuses on how TEK is perceived in contemporary rural communities of the Eastern Cape Province of South Africa, especially regarding the role it plays in the management of natural resources. In spite of their rich history of possessing and applying TEK, the rural communities of the Eastern Cape suffer numerous kinds of environmental challenges, some of which became noticeable during the era of the British colonial state (Beinart 1989, 2003; Palmer and Bennett 2013; Scorer et al. 2019; Murata 2020). The process of making a perception of something is a hermeneutic project which involves using knowledge acquired through our senses to arrive at a judgement that something is X or is not X. Doing so involves deployment of perceptions to assign meaning and significance to various forms of reality. Perceptions are part of people's knowledge because they provide the basis for justifying what we think is true or false (Stacey 1994; Murata 2020).

Several studies across the globe have documented key traditional practices that rural communities implemented in the past and continue to do so albeit at substantially reduced levels (Fabricius et al. 2006; Berkes 2012; Beinart and Brown 2013; Murata et al. 2019). The studies unravel two main common findings that may have a bearing on how the practices might be perceived by different actors in contemporary rural communities. First, that traditional knowledge tends to have limitations in understanding some ecological challenges, especially the ones that are not very physically identifiable. Second, the ontology of traditional resource management practices (for example, application of taboos) does not lend itself readily intelligible to ordinary day-to-day methods of interpreting reality. The findings leave the question on how TEK fares in contemporary rural South Africa, wide open.

Of late, literature has demonstrated a growing ambivalence in the capacity of TEK to provide effective resource management wisdom. The levels of doubt range from partial to total rejection of the importance of leveraging traditional knowledge in resource management initiatives. Total rejection is voiced by authors that argue that TEK is not capable of managing resources hence scientifically trained ecologists must be deployed in the tropics to save nature (Smith and Wishnie 2000; Du Toit et al. 2007). Partial doubt has been registered by authors who argue that although TEK provides useful natural resource management wisdom, it is not capable of identifying and solving environmental challenges that are not physically observable (Fabricius et al. 2006: Costanza 2008; Beinart and Brown 2013; Murata et al. 2019). However, for all its worth, this literature does not address the question on whether the conclusions were reached after finding out that the knowledge system was implemented fully.

Rural communities of South Africa, particularly the Eastern Cape, possess a rich stock of traditional knowledge that dates back to pre-colonial times (Nompozolo 2000; Bank 2002; Denison and Wotshela 2009; Murata 2010). The knowledge includes several practices of environmental resource conservation and use practices which draw heavily on the local belief system (Murata 2020). Nonetheless, since the early twentieth century, following introduction of the colonial state, the rural communities of the Eastern Cape have become environmental hotspots, characterized by heavy land degradation, overgrazing and worsening water shortage (Beinart 1989; Palmer and Bennett 2013), leading to a progressive trend of deagrarianization which began in the closing decades of the twentieth century (Bundy 1988; Andrew and Fox 2004). Palmer and Bennett (2013), as well as Scorer et al. (2019) have explicitly pointed out that land in traditional villages which use the communal land tenure system is more degraded than the one in privately owned, mostly white owned farms. If no scholarly effort is made to find out what actually is happening with traditional ecological knowledge in the communities, the growing argument that contemporary global society should integrate the knowledge with the scientific knowledge system in order to enhance the capacity of contemporary society to solve environmental challenges (Brundtland 1987; Chalmers and Fabricius 2007; Denison and Wotshela 2009; Murata 2010; Ayaa and Waswa 2016; Berkes and Folke 1998; Ngara 2013; Pearce et al. 2015; Sillitoe 2017; Tengo et al. 2007; Convention on Biological Diversity (CBD) 2004; MEA 2005) might soon lose traction.

9.2 Understanding Traditional Ecological Knowledge

Traditional ecological knowledge refers (TEK) to "a cumulative body of knowledge, practice and belief, evolving by adaptive processes and handed down through generations by cultural transmission, about the relationships of living beings (including humans) with one another and with their environment" (Berkes 2012: 7). Traditional ecological knowledge includes the way in which people acquire knowledge (how people get to know that which they know), and the content of the knowledge itself (what is known) (Berkes 2012; LaDuke 1994).

This form of knowledge is regarded as traditional not because it concerns practices or values that were relevant in the past, but because it uses various concepts and precepts of tradition, such as cultural and religious beliefs, taboos and related tribal conventions for its generation, transmission and application (Battiste and

Youngblood 2000). Traditional ways of knowing pertain to the use of a community's culture, its conventions, observances, beliefs and worldviews as the main tools that determine what is known to be real (ontology), and how that which is known gets to be known (epistemology).

Belief in, and respect for tradition, including its key aspects, such as taboos, ancestral spirits and visionary knowledge, form the central cog of sustenance of TEK in a community (Battiste and Youngblood 2000; Berkes 2012; LaDuke 1994). Some scholars have argued that traditional knowledge and its associated practices cannot work hand in hand with post-colonial liberal democracy because the two derive their legitimacy from contradictory sources. While the former emanates from pre-colonial social-political roots, the latter is a successor of and is informed by the colonial nation-state system, which by and large, was and continues to be the pivotal conduit of Western[1] knowledge (Meer and Campbell 2003; Ray 1996).

Studies conducted elsewhere have demonstrated that the enthusiasm that characterised the discourse of indigenous knowledge, especially TEK over the past three decades, which was particularly ignited by the coinage of the sustainable development concept, is undermined by a growing fear that the knowledge system is not embraced, and in some instances, its truth-claims are actively contested. A study by Pieroni et al. (2004) found that some modernising communities in Italy were losing knowledge of traditional medicine. Another study shows that the use and knowledge of plants declined in the Tsimane community of the Bolivian Amazon due to the reality that the local communities integrated into the market economy (Reyes-Garcia et al. 2005). Furthermore, a study of changes in cultural traits among the Amazonian indigenous communities found that due to modernisation, TEK is lost faster in villages that are close to urban centres than in remote ones (Reyes-García et al. 2013).

---

[1] The concept of knowledge should be analysed as a body of knowledge with very transfactual constitutive properties whose existence transcends the geographical boundaries of the Global North.

This is mainly due to the replacement mentality of modernised systems. "Modern" systems believe that "inferior" systems should be replaced. It is however argued that "modern" systems should plug-in into indigenous systems to ensure sustainability and resilience of the systems. Moreover, TEK is contested and hence declining in the Bolivian Amazon and Donana region of Spain due to forces of change including modernisation, education, and integration into the market economy (Gomez-Baggethun and Reyes-Garcia 2013).

More cases of contestation and loss of TEK have been found in Spain (Gómez-Baggethun et al. 2010), Canada (Turner et al. 2000), China (Ahmed et al. 2010), Ecuador (Guest 2002), Vanuatu (McCarter and Gavin 2014) and India, Indonesia and the United Kingdom (Pilgrim et al. 2008). The drivers and processes of loss of TEK are complex and varied, including changes in knowledge systems, conversion from traditional religion to foreign religions, adoption of western education, and state-imposed restrictions in natural resources management (Turner and Turner 2008).

In the South African literature, the question of how TEK is perceived and whether its related management practices are embraced or contested remains largely unaddressed. South African studies have mainly been concerned with investigating how traditional knowledge is used by local communities to manage the natural environment and its services. These studies include Chalmers and Fabricius (2007) who discuss the value of local ecological knowledge in management of ecosystem services on the Wild Coast of South Africa; Denison and Wotshela (2009) and Murata (2010) who investigated indigenous ecological knowledge in managing rainwater and soil resources, and Critchley and Netshikovhela (1998) who investigated traditional practices of soil conservation. Largely, local studies project traditional knowledge as a virtuous system that has a lot to offer in the domain of natural resource conservation and use.

The few South African studies which touch on challenges and weaknesses of traditional knowl-edge systems include Fabricius et al. (2006) who conclude that the knowledge system tends to struggle with explaining causation and coping with foreign-induced environmental changes. Another study is by Beinart and Brown (2013) who found that traditional knowledge system struggles to comprehend and explain minuscule processes such as germ and bacteria infection. Nonetheless, these studies do not look at how traditional knowledge system is perceived. In South Africa. The study that touches on this subject is by Kapfudzaruwa and Sowman (2009). In their study of Water User Associations (WUAs) in South Africa, Kapfudzaruwa and Sowman (2009) found that TEK faces challenges such as systemic exclusion when the state and private institutions design and implement water resource management intervention programs. Instead, traditional ecological knowledge should have been the start of developing a sustainable water governance programme. The authors claim that it is easy to start by what people know, and then integrate it into new ideas for success and sustainability. Nonetheless, Kapfudzaruwa and Sowman (2009) did not do an in-depth investigation of why traditional management practices are sidelined. Consequently, the question of what actually is happening with traditional management practices in rural South Africa continues to remain largely unaddressed.

The contribution of this chapter is twofold. First, it typologises the challenges faced by TEK in contemporary rural Eastern Cape. This helps streamline them according to their core defining features, a good understanding of which can better inform design and implementation of intervention programmes. Second, by making the challenges its central focus, the chapter holds potential to ignite subsequent research interest in the direction not only of celebrating traditional knowledge but investigating if the knowledge system is fully implemented or not. Establishing this knowledge can provide useful groundwork to debates on why in spite of an abundant wealth of ecological knowledge, traditional communities experience high levels of environmental degradation in post-apartheid South Africa.

9.3   Theoretical Inclinations

9.3.1   The Interpretive or Hermeneutic Dimension of Critical Realism

We used critical realism's interpretive or hermeneutic dimension to make sense of the multiple challenges that traditional resource management practices face at the local level. Hermeneutics concerns itself with the complexities of interpreting social phenomena (human action). "Interpretation is a ubiquitous activity, unfolding whenever humans aspire to grasp whatever *interpretanda* they deem significant" (Mantzavinos 2020: no page number). There are two levels of hermeneutics; ontological hermeneutics and epistemological hermeneutics (Mantzavinos 2020). While ontological hermeneutics deals with interpretation of being (existentialism), epistemological hermeneutics concerns itself with ascription of meaning to reality (being). In this study we used the latter because what is at stake is not the existential nature of traditional management practices (e.g. what taboos are); rather it is whether or not they are interpreted as constituting resource management techniques.

The challenges facing TEK in rural communities include divergence of worldviews and religious beliefs that play out between holders of TEK, who are the main actors in the implementation of these practices, and non-traditionally minded people. The latter group largely comprises Christian converts and people with fairly high levels of formal education, from secondary to tertiary. The main contentious issues revolve around interpretation and meanings that are ascribed to traditional management practices such as taboos, as well as the role that supernatural forces including ancestors play in environmental management. These interpretations and their associated truth-claims tend to be informed by and reproduce worldview polarities that manifest between the local traditional religion and Christianity. While the former acknowledges the active role of ancestors in protecting the welfare of humans and the integrity of nature, the local version of Christianity does not recognise this.

Furthermore, the contestations over the place of traditional practices in environmental management tend to coalesce around the differences (perceived or otherwise) between formal education and indigenous knowledge system (IKS) in so far as the two strive to explain the world, including drivers and processes of ecological change.

The hermeneutic dimension of critical realism argues that all social phenomena including objects, processes and abstracts "are intrinsically meaningful, and hence meaning is not only externally descriptive of them but constitutive of them" (Sayer 2000: 17). In light of this, all social phenomena including management practices inherently carry meanings, and these meanings are not mere social constructs, but possess substantive existential properties with objective ontology. Nonetheless, meaning cannot be known unless it is accessed, interpreted and communicated. The process of doing this is influenced by the interpreter and communicator's social properties including their worldview, religious beliefs and epistemic orientation (Bhaskar 2016; Danermark et al. 2002). So depending on their socialisation, different interpreters may see and communicate equally different meanings of the same social phenomenon. These different meanings can be a constraint to successful interpersonal and intercommunity dialogue over certain realities of interest.

Although such realities (e.g. traditional management practices) exist in an objective state, that is they are what they are regardless of our socially influenced interpretations of them, the meanings we ascribe to them can only be understood, and not measured (Bhaskar 2016; Sayer 2000). Consequently, there is always a hermeneutic element in social life because in order for two or more actors to get into a dialogue about a particular reality, they have to understand each other's interpretations.

Social actors are not only confronted with the task of having to interpret the reality communicated to them by their counterparts (e.g. that they should not kill certain species of frogs because doing so can upset ancestors who in turn may causally lead to drying up of a spring), they as well need to enter the hermeneutic circles of the communicators. In this regard, non-holders of TEK do not only need to interpret the taboo that says killing *Gyrinidae* (*inkwili*) can causally lead to drying up of a water resource, but they also need to enter the epistemic world of the individual who communicates this. This is a double hermeneutic; a burden of interpretation that does not apply to scientific actors (Sayer 2000). Scientific actors whose hermeneutic remit is primarily to make sense of physical objects (e.g. soil, rocks and trees) do not suffer the burden of double hermeneutics (Sayer 2000).

Bhaskar (2016) explains this hermeneutic complexity by disaggregating reality into transitive and intransitive dimensions. The intransitive dimension is the social phenomenon (e.g. making use a certain tree species a taboo) while the transitive dimension is constituted by our interpretation of what that phenomenon means; that is whether we regard the taboo as a resource management practice or we dismiss it as a mere fable. The transitive dimension may change or vary from one social actor to the other, but the phenomenon (intransitive dimension) remains what it is (a taboo) (Bhaskar 2016).

Social meanings are related to and significantly shaped by the context in which the social phenomena unfold. Contextual factors that influence what we make of social phenomena include cultural, economic (lifestyle), social (the kind, and levels of education acquired) and political (rules, laws and policies) realms. What may constitute environmental management to an individual with formal education training may not be similar to what a holder of traditional knowledge thinks. And what may constitute a fact to the latter can be interpreted as a fable by the former.

9.3.2 Critical Realist Dimension of Dynamism

The second critical realist dimension used in this chapter is that reality, including events and processes, occurs in a dynamic open system that cannot be directly controlled (Bhaskar 2016;

Danermark et al. 2002; Sayer 2000). This contrasts with the positivist approach that studies phenomena in closed laboratory-controlled environments (Wynn Jr and Williams 2012). Social realities, such as application and transmission of TEK, cannot be contained and controlled in a laboratory-like experiment environment; they depend on a multitude of changing conditions, as communities evolve over different eras (Berkes 2012). Attendant conditions *inter alia*, social-cultural, political, economic, although not in a strictly causal way, possess generative power to either enable or constrain the occurrence of phenomena such as people's trust in and use of taboos in resource management, and their belief that ancestors interact with the living in ways that can change the health and functioning of the natural environment. Shifts and developments, especially in the domain of political administration and livelihoods that have taken place in post-apartheid South Africa (See section on study context) have the potential to influence how local communities perceive traditional knowledge and its associated practices.

The boundaries within which social phenomena unfold are typically dynamic; they change over time. These changes make it difficult to assume, as several researchers and international organisations do, that traditional management practices that were applied in the past, in different contextual environments, are still applicable in the contemporary traditional communities (Wynn Jr and Williams 2012). Precisely because of its acknowledgement of dynamism, a study with a critical realist approach shifts its focus from the positivist concern with repeatability of events to tendencies of realities to occur within a particular contextual setup and time (Wynn Jr and Williams 2012). Thus, changing conditions within communities may fundamentally influence people's perceptions of different traditional practices of resources management. A vast amount of literature has demonstrated this point in many study cases (Ahmed et al. 2010; Guest 2002; Gomez-Baggethun and Reyes-Garcia 2013; McCarter and Gavin 2014; Reyes-Garcia et al. 2005; Reyes-García et al. 2013; Turner et al. 2000).

9.4 Context of the Study

This section is presented in a way that departs from the conventional approach of describing a study site. Instead of focusing on biophysical, social and economic characteristics of the study sites specifically, the section presents a wider national social, economic and political overview. This less conventional approach of describing study sites was necessitated by the realisation that factors that influence how local people perceive the role of traditional practices in environmental management might be of a national scale.

This study was conducted at three remote isiXhosa-speaking villages of Colana, Gogela and Nozitshena in the Eastern Cape Province of South Africa. Upon achieving democracy in 1994, South Africa adopted a national constitution built on liberal principles that include equality of persons and gender, human rights and governance through democratically elected representatives (Republic of South Africa (RSA) 1996). While the constitution provides for leadership by elected government officials, in sections 211–212, it recognises the role of traditional leaders to continue to supervise the use of indigenous knowledge, customs and customary law under its regulation (RSA 1996; Bank and Southall 1996). The constitution also provides that traditional leaders serve as *ex officio* members in local government and advisers at provincial and national government levels (Bank and Southall 1996).

The state determines the scope and nature of the key responsibilities and power of traditional leaders in maintaining community culture and customs, as well as presiding over traditional rural courts that use customary law (Republic of South Africa (RSA) 1998). The constitution created three spheres of government—national, provincial and municipal—and provides that municipalities be formed in all areas of South Africa, including in rural villages headed by traditional leaders (Republic of South Africa 1996). The state and traditional leadership systems operate quite differently, although some scholars have advanced the argument that the latter can be used to complement the former (Sklar 1986). The dis-

tinction between the two led Bank and Southall (1996) to describe post-apartheid leadership arrangements as a case of several traditional states behind the state. Ntsebeza (2004) argues that post-apartheid South Africa's constitutional position of mixing liberal democratic institutions with unelected traditional leadership is contradictory. This contradiction is a source of confusion for traditional leaders and rural people, and its influence has spilt into the domain of natural resources management.

Traditional leaders are not sure about the scope of their power in community matters such as justice and law, land administration and control over natural resources including forests, pasture and water resources. Prior to democracy (1994) these community matters were presided over by traditional leaders, albeit with indirect state intervention (Ntsebeza 2004). Faced with the introduction of elected municipal officials including ward councillors, rural people are confused with regard to which structure of leadership to pay allegiance to: traditional leaders or elected state officials (interview with Mr. Mdletshe, 10 October 2018). In some villages, traditional leaders have withdrawn from all community administrative activities, including management of natural resources, and have left these matters to ward councillors and civic organisations (Bennett 2013).

Furthermore, the space of customary law in democratic South Africa is not clear. While the national constitution recognises customary law and provides that traditional leaders can apply it in rural communities, section 211(2) of the constitution gives the legislature power to repeal any customary law and replace it by civil law if that law is found not to be in line with principles of liberal democracy (Kapfudzaruwa and Sowman 2009; RSA 1996).

As forces of social change such as formal education, Christianity, monetization of rural livelihoods and material accumulation gravitate to the countryside, rural communities are caught up in intra-household and intra-community contestations over identity formation. The contestations are fought between two cultural formations; the *qaba* (those resisting modernisation) and the *gqoboka* (those modernised, including Christians and formally educated) (Bank 2002). While the former emphasise the original traditional amaXhosa cultural practices, including propitiating ancestors, holding traditional ceremonies and traditional knowledge system, the latter adopted Western culture. They embraced Christianity and worship the Holy Spirit, they emphasise formal education as the sole source of wisdom. These are the political and social contexts within which the application of TEK takes place in post-apartheid rural South Africa.

9.5 Methods and Approach

The study used a qualitative case study approach in order to give respondents space to narrate their knowledge, experience and perceptions about the study theme without the restriction of having to choose from a pre-determined set of responses, as is the case in quantitative surveys. It was necessary to use this methodological approach because it situates our understanding of how TEK fares in the local communities in the perceptions of the actors on the ground. Moreover, the qualitative approach makes for a good complement of critical realism used as the theoretical lens because both hold that knowledge can be understood and not measured (Bhaskar 2016; Merriam and Tisdell 2016; Mantzavinos 2020).

A snowball sampling technique was used to recruit a total of 83 informants: 35 from Colana, 33 from Gogela and 17 from Nozitshena villages. We were interested in recruiting the elderly women and men of 60 years and older. In addition to the advanced age, these elderly community members needed to belong to families who had a long history of living in these case villages. Such history should stretch for at least two generations. These criteria were important in that they enabled us to pick informants who had lived through many decades of the histories of their villages. Such informants were better positioned than the young ones to compare how TEK was applied in the past with the present situation.

Working with two local field assistants at each case village, we administered open-ended interview sessions in the local isiXhosa language. The sessions lasted between 50 minutes and 1.5 hours. We digitally recorded the interviews using a cell phone and tape recorder. Although the questions were open-ended, we concentrated on getting information on the following sub-thematic areas: first, traditional practices that were actually used to manage natural resources at the local communities. Specifically, questions on this matter sought to find out if traditional concepts such as taboos, sacredness and social conventions were used. This question was asked because traditions may vary from community to community. In fact, different beliefs may lead to different ways of practising TEK. Second, perceived and empirically experienced challenges that confront efforts to implement traditional management practices at the three villages. Questions on this sub-theme were particularly directed at eliciting respondents' perceptions of their fellow community members, and even government officials towards use of TEK as a source of resource management wisdom.

9.5.1 Data Analysis

The process of data analysis was guided by a critical realist inferential technique called abduction. In critical realism, the concept of inference is not used to denote the conventional notion of logical derivations (through induction and or deduction), but rather, it refers to organised ways of reasoning, thinking and arguing in a wider sense (Danermark et al. 2002). Unlike its meta-theoretical counterparts, such as positivism, which are interested in empirical generalisations through induction and deduction, critical realism is interested in reaching transfactual or theoretical generalisations through applying abductive reasoning (Bhaskar 2016; Danermark et al. 2002).

Abduction involves studying empirically concrete events and use them to understand underlying, unobservable structures (Danermark et al.

2002). The process of doing this involves re-description or re-contextualisation of events, mostly in terms of a causal mechanism or process that serves to explain the state, condition or happening in a causal manner. The abductive analytical technique provides a good fit to the concepts of hermeneutics and dynamism because re-description of phenomena involves interpretation and contextualisation in, *inter alia*, time and space.

While deduction proves that something must be in a certain way, abduction shows how something might have become what it is. Abductive reasoning achieves this by proceeding from conceiving a concrete event to a deeper conception of it. This means to "observe, describe, interpret and explain something within the frame of a new context" (Danermark et al. 2002: 91). When analysing data on challenges that traditional practices of resource management confront in local communities, we related the individual testimonies to wider and structural phenomena such as political, social or economic forces.

After transcribing the data, we used hand-coding for analysis. Because we wanted to move from the level of observable events to transfactual meanings, the process of data analysis involved four iterative stages. In the first stage we analysed three transcripts, one from each village case. We wrote down all narratives that made reference to challenges of using TEK in managing natural resources These challenges became our first codes. After that we analysed all the transcripts, grouping all challenge-related narratives under these codes. We created new codes in cases where we found new challenges. The second stage involved re-reading the codes in order to identify similarities and differences between them. In this process we grouped together all codes that were similar and left separate all those that were different. For example, codes such as 'children do not respect traditional beliefs any more' and 'children do not understand our culture' were grouped together as one code.

Stage three involved the search for transfactual meanings in which we looked for deeper

structural mechanisms that could underlie each code. Transfactual coding was guided by a typical abductive question: what is it that is general in these individual narratives of empirical phenomena? For instance, in the codes such as 'the educated disregard use of taboos as mere fable' and the one that says 'people regard ancestors as demons', we found that worldview was the underlying general mechanism here because it is a challenge that emanates from the reality that certain people see the world differently from the way holders of TEK do. We got three main transfactual codes at this stage of analysis: worldview, institutional, and lifestyle. These codes enabled us to see the challenges as general and structural rather than individual, isolated realities.

The fourth and last stage involved splitting the transfactual codes into more specific sectors. For example, the concept of worldview is wide, hence leaving analysis at that level might be unhelpful. Consequently, we then looked for more specific issues or aspects of these transfactual challenges. As a result, we found three sub-codes under world view, two under institutional, and one under lifestyle.

The core analytical property of abduction is its ability to describe individual concrete events as both individual phenomena and as manifestation of, or part of general structures (Danermark et al. 2002). It was helpful in this study because we wanted to find meanings about the contestations and resistances that confront TEK in contemporary South Africa. Although they are easily observable, such contestations do not often betray their linkages with structural forces such as changes in governance, belief and educational systems. Studies that ignore these underlying linkages may not be able to streamline the challenges in ways that facilitate design of sector-specific intervention. In a way, such studies may not be able to show the sectors (whether national governance policy, relationship between religions, or the relationship between traditional knowledge system and Western education system) that need intervention in order to solve these challenges.

9.6 Findings and Discussions

9.6.1 Worldview Challenges Confronting Traditional Management Practices

A worldview is a cognitive orientation that provides a framework within which individuals or communities interpret reality; that is how they understand and perceive that which exists or happens (Collins English Dictionary 2005; Oxford Advanced Learner's Dictionary 2015). This cognitive orientation exerts an influence on what individuals or communities believe or hold to be true knowledge, take as real or mythical, and consider as important or useless in varied domains of life, including management of the environment. The worldview challenges found at the three communities are subdivided into a diminishing sense of nation, disrespect for taboos and social conventions.

9.6.1.1 Diminishing Sense of Nation

All respondents said the most important principle that enabled previous generations in the past to manage resources using traditional practices was respect for the nation. Traditional people's concept of the nation is different from that of the modern state system which denotes a collection of all the people and their lands which live under the same government (Oxford Advanced Learner's Dictionary 2015). The respondents spoke of the nation (*isizwe*) as their space of territorial jurisdiction, its resources, and the dead and living people. The sense of nationhood emanates from a feeling that they belong to and enjoy jurisdictional power over the land of the community and its resources. It is the symbiotic interaction and interdependence between the living, the dead, and the natural environment that make up the concept of *isizwe* (nation).

The majority of respondents (70%) said this complex circle of interaction among people, natural environment and the dead has been significantly broken by the intervention of the state in the space of resources management at local community level. The state has imposed its jurisdiction over local resources in both ownership and

management through the Department of Environmental Affairs which owns and manages natural resources including forests, the Department of Water and Sanitation which owns and manages unallocated water resources in these communal villages, and the Department of Agriculture and Land Reform which is the owner of all communal land resources.

Consequently, the interrelationships between local people and their natural environment is no longer as robust as it was in the past. Some of the resources management rules that the state has imposed, such as forbidding local communities to harvest thatch grass on roadsides, force local people to cut the grass at night. This practice of 'harvesting by the night' is in a way a form of resistance to state interference. The sense of environmental morality that connected local people and nature in the past has been blocked by the imposing image of the state. Although local people continue to see the same image of forests, grasslands and species of wild animals they saw in the past, at least at the empirical level of reality, the sense of moral attachment and ethical responsibility to care for nature has been replaced by fear and hate of the punitive, impersonal state environmental laws and regulations. One of the state environmental laws that have displaced traditional management practices, pertains to burning of veld fires for management of pasture resources. In the past, local communities set fire on grazing lands at set times of the local ecological calendar to both induce renewal of grass and control tick-borne diseases. However, the state has taken over the role of setting veld fires, in the process making it a criminal offence for local people to burn the veld without permission from the state. This is clearly a case of the state (modern system) refusing to "plug-in" to an existing system and thus causing significant disequilibrium and resistance which could have been avoided.

The loss of the concept of *isizwe* among local people can be understood by appreciating the effects that context can have in shaping our understanding of reality. The arrival of the state in the local social-ecological space provided a different context which re-configured the relations between people and nature. The state-dominated context causally changed the roles and responsibilities of local people towards nature from being owners with a morally sanctioned duty of care to poachers who have to be policed lest their actions destroy nature. This context-induced shift of roles bears a hermeneutic implication which further constrains usage of traditional practices at the three communities. Because the state has designated them as environmental villains to be policed and fined, the local people's interpretation of what nature is has shifted from being a communal asset that needs their care, to being a state resource that should be accessed using 'harvesting by night' strategies.

The significance of the concept of *isizwe* in sustaining and practising traditional management techniques may not be appreciated by officials who lack experiential knowledge of the relationship between traditional communities and their local environments. As demonstrated above, traditional people understand nature not as the elements of resources that exist on the landscape, but as the interactions and interrelationships between these elements and society. In this regard, state officials have broken the interrelationship between local people and their environments. Thus, the role of state officials in management of local resources needs to be renegotiated in order to take account of traditional ecological knowledge and the role of local communities in resource management.

9.6.1.2 Disrespect for Taboos and Social Conventions

Traditional ecological knowledge is informed and heavily influenced by a cognitive orientation that sees supernatural forces as part of the ecological world, ones who possess power to generate events that can lead to either negative or positive effects (Berkes 2012). Respondents reported that this was the worldview of past generations in their communities. In the past, the communities of Colana, Gogela and Nozitshena used taboos to protect useful and high-demand natural resources from degradation and over-exploitation.

The resources that were assigned taboo values were mainly those that were very important to community life and could easily become scarce. Such resources included spring water on which communities depended for drinking, brewing beer, and cooking food. Spring water was protected through a belief system in which biotic species, including *Potamonautes sidneyi* (*unonkala*), *Xenopus laevis* (*noplata*) and *Gyrinidae* (*inkwili*), were responsible for protecting springs from drying up. The species were accorded a special status in the communities, making it taboo to kill them, in the belief that killing them or chasing them from the springs would cause the springs to dry up and lead to a shortage of a critical ecosystem service; drinking water.

Respondents reported that, in the past, their forebears deeply respected the relationship between nature and people. Some natural resources, such as snakes and chameleons were assigned totemic status, and by so doing, were protected from being killed wantonly. Examples include *Pseudaspis cana* (*majola* snake) and *Chamaeleonidae* (*sigcilikishi*) which are totemic animals to some amaXhosa clans. These natural resources were believed (and still are, but to a lesser extent) to provide cultural ecosystem services which include media of communication between the living and ancestors, bringing luck and good health to the living and general protection of families and clans against their enemies, especially evil spirits. When people of these relevant clans, for instance, killed *Pseudaspis cana* by accident, they apologised to their ancestors and dug a hole in which to bury it with dignity.

According to a local legend, a woman who married into the Majola clan (a clan named after *Pseudaspis cana*) could not pull up her skirt when crossing a river in which the snake was believed to inhabit. The service that *Pseudaspis cana* provided was to protect the river from drying up and prevented waterborne accidents, such as drowning and of floods washing away people. As a result, the snake was highly respected among the local people. They would rather wet the edges of their dresses because *Pseudaspis cana* was in that water. Pulling up their dresses would traditionally be seen as a sign of disrespect for the snake and was believed to invite dire consequences, such as drowning and destructive floods. All respondents reported that, currently, some of the local people do not respect theses traditions anymore. They wantonly kill the *Pseudaspis cana* snake, and violate the taboos. The perpetrators contest the role of taboos in facilitating and regulating human-nature relations. As a result, said the respondents, many rivers are drying up and drowning accidents are frequently happening. There were two cases of drowning at Gogela village in 2018 (during data collection for this work) and respondents interpreted the incidents as results of ancestral anger. Respondents also reported that, over the previous 20 years following democracy, several rivers had dried up because some of the local people disrespect tradition. These include three rivers at Colana, two rivers and a stream at Gogela, and two perennial streams at Nozitshena which became ephemeral.

According to the local people's TEK, it is also taboo to kill a *Danaus plexippus (iphelandle)*. *Danaus plexippus* are a critical part of the traditional agro-ecological calendar. They signal coming of rainwater, thus they provide an important ecosystem service. Traditionally, when *Danaus plexippus* were seen flying in a group moving from east to west, local people knew that rain was coming shortly. This helped farmers prepare for tilling the soil and planting. Killing *Danaus plexippus* was made a taboo in the belief that if one killed it, their livestock would not multiply, but would start dying, one by one. Nonetheless, respondents reported that now some of the local people disobey this taboo. One major consequence of this is that local farmers have lost an important ecosystem service, especially given the rainfall variability associated with climate change.

Furthermore, the taboo system was used for management of forest resources, especially trees. Households were forbidden from using taboo-trees, both as fuel wood or construction timber. Most of these trees provide a range of important ecosystem services that included cultural, aesthetic, and provisioning. The cultural service included use in traditional rituals. Aesthetic services included making the landscape beautiful, simply good to look at. Lastly, provisioning services included goods such as wild fruits and medicinal plants.

Some of the taboo trees and the negative consequences that were believed to befall people who violated the taboos include:

(i) *Phytolacca dioica*. This is an aggressive alien invasive species and is locally known as *Sdungamzi* (causer of family trouble). It originated in South America and arrived in South Africa in the twentieth century. This tree may not be harvested or used for any purpose within the homestead. It is believed that families that violate the taboo suffer incessant intra-family conflicts that can result in total disintegration.

(ii) *Rapanea melanophloeos*. This tree is locally called *Sqoni*. It is taboo to use *Rapanea melanophloeos* as fuel wood or construction timber. It is believed that the fire made from this tree causes the testicles of male family members to swell and heavy menstrual events for women.

(iii) *Gymnosporia nemorosa*. This tree is locally called *Mbangandlala*. It is taboo to use it as fuel wood or construction timber. The word *Mbangandlala* literally translates as 'one that causes hunger or poverty'. It is believed that inhaling smoke or absorbing heat from the tree's fire causes hunger or poverty through a number of misfortunes, including losing one's job or having family members lose their jobs wherever they are employed. Once this happens, their families suffer from hunger or poverty.

(iv) *Gardenia thunbergia*, locally called *Sendelenja*. The term '*Sendelenja*' literally translates as 'a testicle of a dog'. According to local tradition, it is taboo to use the tree as fuel wood because it is believed that the fire from it causes men's testicles to swell and become painful.

Respondents said that all these taboo-trees are harvested in the current times and used for various domestic needs, including fuel wood and construction timber. When asked why people harvest taboo-trees, respondents said the perpetrators argue that taboos are simply a fable that should not be taken seriously. They argue that cutting a tree cannot cause painful menstruation in women, or a family member to lose a job. The perpetrators argue that there is no link between these trees and the biological system of a woman, nor can the trees influence the relationships between working family members and their employers.

Disrespect for tradition, especially taboos can be looked at as a hermeneutical issue that is playing out as liability or constraint to the implementation of traditional management practices. The contestations are not about opposing existential knowledge of taboos; whether some people do not understand that taboos are taboos. Instead, it is a hermeneutic question emanating from shared differences in making meaning of the role that taboos can play in resource management. This interpretational difference is shaped by, and manifests fundamental epistemic dissimilarities between TEK and Western education which is largely scientific by way of its methods of enquiry and verification of truth-claims, as well as ontological position (what is reality and what is not) (Bhaskar 2016; Berkes 2012). The interpretations held by holders of TEK in which they ascribe management role to taboos is based on belief, albeit not necessarily a fable. Unless the other social actors on the other side of the epistemic divide can successfully enter a double hermeneutic in which they do not only strive to make sense of the taboo, but also make sense of what is going on in the cognitive processes of the traditionalists, this interpretational challenge will never get to be resolved. So, in local communities what is constraining traditional management of resources is not that the knowledge system uses taboos, but it is because the power of taboos in managing resources is not understood by some social actors.

Furthermore, the dimension of dynamism can help explain why use of taboos in resource management is contested in the present times, but was embraced in the past. Acceptance of the claim that taboos are management tools is based on a worldview which sees supernatural forces as guardians of both society and the natural environment. This worldview was very strong and prevalent in the past, hence taboos were embraced in resource management practices. However, worldviews, just like their associated social realities including culture, are dynamic; they evolve and adapt due to societal changes. When used as an analytical tool, the critical realist dimension of dynamism can help researchers and policymakers to appreciate that TEK gets affected by changes of societies in which it is practiced.

9.6.1.3 Disrespect for Legends

In addition to taboos, legends were used, and still continue to be used but to a much smaller extent, as resource management tools. There is a well-known legend of a big marshy, perennial wetland locally named Entombini (the lady's place) that existed at Gogela village over a long time before inception of democratic government in South Africa. The wetland was given this name because people frequently saw a mysterious, beautiful Caucasian lady there. Community members occasionally saw the lady hang her laundry on surrounding trees and tall grass. The wetland was the source of a stream which supplied clean drinking water to the village. It was believed that the lady was the owner of the wetland, responsible for managing it. Anthropogenic activities such as pollution and misuse were perceived to be the major threats to the ecological integrity of the wetland, which could ultimately upset the lady. It was feared that if the lady got upset, she might leave, leading to the drying up of wetland.

Consequently, elders of the village set rules to regulate how people interacted with the Entombini wetland. These included prohibiting activities such as bathing in the wetland, watering livestock in the wetland and urinating, having sex, as well as defecating near the wetland. It was believed that all these were dirty activities that could upset the lady. Nonetheless, starting from the late 1990s, some local people started to flout the rules, doing all sorts of things at the wetland. Consequently, the lady left and the wetland began to progressively shrink into small sizes, leading to complete drying up by 2002.

The worldview that shaped and gave meaning to all these traditional practices is waning. Several influential people, including university graduates, church leaders and technocrats, in the local communities do not perceive supernatural forces as part of the ecological realm. Among them, the extreme ones, even completely reject existence of the supernatural forces. Respondents blamed the infiltration of Western education and Christianity for this shift in worldview. There is a growing general negativism against traditional management practices, among educated community members who perceive them as signs of *ubuqaba* (uneducatedness) and backwardness. The negativism is compounded by converts of Christianity who perceive traditional practices as belonging in the realm of demons and darkness.

As a result, efforts mainly by the elderly to enforce rules, taboos and beliefs of traditional resource management are perceived as old-age beliefs devoid of practical value. The *gqoboka* (people with Western education) challenge traditional people to prove how taboos and other facets of tradition can actually affect the functioning of the natural environment. However, holders of TEK find it difficult to prove this because traditional practices and beliefs cannot be subjected to experimental testing. Nor can their ways of working be empirically observed.

Sacred biotic species that, in the past, were believed to possess power to manage water resources including *Potamonautes sidneyi*, *Xenopus laevis*, *Danaus plexippus* and *Gyrinidae* are wantonly killed by the current generations without fear of consequences. Furthermore, sacred trees such as *Gymnosporia nemorosa*, *Rapanea melanophloeos*, *Gardenia thunbergia* and *Phytolacca dioica* are harvested for domestic use by several households. Respondents argued that this had led to environmental challenges such as drying up of springs, late rains, storms and heavy snow.

These changes in how people interpret the world can be explained by the realist disaggregative conceptualization of reality as a layered ontology comprising the intransitive and the transitive dimensions. Every era or generation (albeit noting that generations are not homogenous realities) in human history has its own interpretations (transitive dimension) of social phenomena, objects and concepts. Such interpretations depend on each generation's key social properties such as education, worldview, culture and religion. Interpretations change with the changing of the attendant key social properties and the passing of generations, but these changes do not mean that the intransitive dimension, that is the realities to which they ascribe meaning, such as traditional management practices have changed (Bhaskar 2016). The reality that taboo-species, legends and sacred resources were perceived as management tools by past generations, but now dismissed as useless fictions by current generations does not mean they have changed.

Moreover, all knowledge is interpretive and not representational of the things about which it is (Bhaskar 2016; Sayer 2000). Whether it is scientific knowledge or TEK, the best it can ever do is to postulate truth-claims that are capable of describing reality better than any other available rival claims during that time period. This means all knowledge is fallible. Traditional ecological knowledge is, of course, not the silver bullet. However, the claim that TEK cannot perform certain functions should be taken as an opportunity to further develop it, rather than dismiss and reject it. All knowledge is limited and hence fallible (Bhaskar 2016).

9.6.1.4 Adoption of Christianity and Western Education

Religion and belief play a crucial role in the generation and implementation of traditional management practices. Historically, the communities of Colana, Gogela and Nozitshena believed in ancestors whom they regarded as the link between them and *uThixo* (the creator). Through dreams, spirit media, visions and symbolic events such as seeing a black cat cross one's way, ancestors communicate knowledge and warnings to the living about matters of life, including society-environment relations. Ancestral messages included where and how to harvest medicinal plants, wild animals, fruits and honey. The messages also served as early warning signs of impending ecological disasters such as drought events, thunderstorms, wind and hailstorms. However, it has become difficult these days for holders of TEK to tell their fellow community members about knowledge from dreams because these media of knowledge construction are chastised as demonic and superstitious. The version of Christianity preached in some churches in the local communities argues that there is nothing called 'ancestors'; when a person dies, they are dead. The dead cannot come back and talk to the living. This local version teaches its followers that what they think are ancestors are in actual fact demons, hence they must pray hard and chase them away every time they come to their dreams. Consequently, adoption of Christianity significantly undermines TEK which takes seriously the role of ancestors in generating and transmitting ecological knowledge.

The changes in worldviews have as well led to misinterpretation of some of the traditional management practices by non-traditionalists. For instance, non-traditionalists confuse the traditional practice of rainwater management for rainmaking. Rainwater management involves key traditionalists such as *amaxhwele* (traditional healers), community tribal leaders, and *izanuse* (spirit media) who communicate with spirits from the ancestral world when there are rainwater challenges such as delayed rains, stormy rains which destroy crops and houses and thunderous rains which pose danger to human life. These people were consulted by the community to ask ancestors to solve these challenges. In some cases the consultation could be accompanied by performance of traditional ceremonies which involved brewing of *umqombothi* (traditional beer) and slaughtering of livestock, especially goats and cattle.

However, from the inception of democratic governance in 1994, local people have progressively abandoned these traditional practices of rainwater management. All the respondents said the *gqoboka* contest this practice and argue that no one can make rain, an argument that is often made in an effort to discredit the power of traditional people in respect of rainwater management.

The word 'rainmaker' itself is problematic and contributes significantly to this contestation. Dictionary definitions refer to a rainmaker as a person with power to make rain (The Collins English Dictionary 2005; Oxford Advanced Learner's Dictionary 2015). The translated name, 'rainmaker', gives the impression that traditional people claim to possess power to make rain. This has earned holders of TEK a barrage of attacks from people of scientific knowledge (Guthiga and Newsham 2011). But the criticism is misplaced because it emanates from a misinterpretation of what traditionalists do when they practise rainwater management. Rainwater managers are locally called *izangoma zemvula* (doctors of rainwater), not *abenzi bemvula* (makers of rainwater) as the dictionary translation and critiques of traditional rain management say. The individuals who are well known for managing rainwater are locally given praise names such as *Izanemvula* (he or she who comes with rainwater). For this reason, it is clear that tendencies to call traditional people that deal with rainwater challenges as 'rainmakers' are misplaced and cause a fundamental hermeneutic confusion.

Critiques of TEK capitalise on this widely-shared hermeneutical mistake and charge that if holders of TEK possess power to make rain, why did their villages suffer the same way other communities did in the 1973/74, 1983/84 and 1991/1992[2] drought events? However, the truth is that the rituals were never used to create or make rainwater. The mistranslation has endured among scholars to this day (e.g. Guthiga and Newsham 2011; Jackson et al. 2005).

The actual role of *izangoma zemvula*, is to control and manipulate how and when rainwater comes so that communities could maximise their chances of benefiting from it. One respondent at Colana and another at Gogela clarified this confusion: by saying that if there was no rain in the sky in a particular calendar year, no rainwater would fall. However, if rainwater was there in the skies, rituals by rainwater managers could help

[2] Bundy (1988) also acknowledges that these were drought years in South Africa.

control when and how it falls in order to enhance the well-being of community members.

The problem of misinterpreting TEK generally, and its related management practices in particular, constrains effective communication between holders of TEK and non-traditionalists, especially the young generation. All the respondents stated that when the educated are told not to do certain things because they are taboo, such as having sexual intercourse in sacred forests or bushes, they do not take this seriously. In line with their training in Western education system whose methods of knowledge construction are largely informed by the concept of logical empiricism, the educated challenge holders of TEK by asking what will happen if they break the taboos, for instance, by swimming in sacred pools and killing taboo-species. Consequently, elderly people have nicknamed the educated young generation as *bonokontoni* (those who always ask, what will happen if we do it?). The practice of questioning instructions or wisdom is not encouraged in TEK; holders of TEK regard questioning as disrespectful. This point is supported by Berkes (2012) who characterises TEK as a belief-knowledge-practice complex. According to this characterisation, TEK is [true] knowledge by virtue of the fact that it is believed to be so. And by believing it to be [true] knowledge, people put TEK into practice in circumstances such as traditional management of ecosystem services.

Furthermore, the inception of Christianity at the three local communities induced a contextual change which impacts how people look at traditional knowledge. This brought new ways of understanding the ecological realm that are different from what was previously known. Christianity brought a belief system that countered the local religion's respect for and belief in ancestors. The impact of Christianity's counter belief system to local people's perception of traditional management practices cannot be overemphasised. Traditional management techniques such as legends, sacredness and taboos base their legitimacy on the belief held in traditional religion that ancestral spirits can dwell in ecological species and use them as vehicles to reach out to the wider social and ecological realms either to provide protection or to administer punishment. By countering this belief, Christian-believing people has basically gone a long way to removing the launching pad of traditional management practices.

9.6.2 Institutional Challenges Confronting Traditional Management Practices

The concept of institutions refers to the rules of the game, including the game itself (North 1990). Drawing on this conception, institutional challenges pertain to constraints that efforts to implement traditional management practices face as a consequence of the influence of the state system which manifests through, *inter alia*, legislative and organisational provisions. There is a great deal of disharmony playing out at the community level, between the democratic state system and traditional community systems. Such challenges include the existence of legal pluralism which manifests in an adverse co-existence of civil law and customary law (Murata et al. 2022), dethroning of traditional leaders from leadership of community affairs and state overreach in the domestic space.

9.6.2.1 Legal Pluralism and Dethroning of Traditional Leaders from Leading Community Affairs

Respondents reported that the introduction of legal pluralism in the post-apartheid South Africa posed massive challenges to efforts to transmit TEK; including its related resource management practices. The concept of legal pluralism refers to the co-existence and use of more than one legal system at any level of social organisation (Murata et al. 2022; Von Benda-Beckman 2002). From pre-colonial to colonial[3] times, albeit to a lesser

---

[3]Colonial administration in this region began in 1840 with conquest and subjugation of the then independent tribal communities. From then till the mid-twentieth century, the time of the inception of apartheid, the British colonial government implemented indirect rule in the region. During apartheid, the government set this region as a

extent during the latter, local communities used customary law as the legal framework to set standards of actions and behaviour. From 1994, the democratic state has vigorously implemented civil law across the country, including rural communities, which had been enclaves of customary law over the previous eras. Civil law is administered by institutions of the state, including the local municipality and government departments. These two legal orders apply in all facets of community life, including the domain of resource management.

Several respondents (73%) reported that municipal officials, as well as officials from the Department of Agriculture and Land Reform and the Department of Environmental Affairs, introduced resource management rules that are different from the local traditional ones. Historically, the communities used customary law to empower traditional leaders to implement resource management practices, including enforcing taboos and observance of sacred resources. Such practices included decisions about when and how to burn rangelands, and how to manage harvesting of rare and important tree species such as *Ptaeroxylon obliquum* (*mthathi*) and *Salix mucronata* (*mngcunube*). This role has been shifted from traditional leaders to government officials, in the process displacing traditional management practices.

Traditional ecological knowledge depends on customary law for regulation of the ways in which communities interact with the natural environment such as harvesting of resources and observance of the rules associated with different times of the traditional agro-ecological calendar. Over the historical past, customary law put in place a plethora of socio-ecological laws. For instance, there are times of the year when working in an arable field when the sun is overhead constitutes a customary offence because doing so interferes with processes of formation of rainwater in the sky. Customary law provides that the

land and related natural resources in the local communities fall under the custodianship of local traditional leadership. Accordingly, the traditional leaders, in consultation with their people, have historically been, at least before the beginning of state overreach, the local environmental managers with institutional power to enforce management rules.

In contrast, the civil legal order provides that the state is the owner of all communal lands and their resources (Cousins 2008). It vests all responsibilities and power of natural resource management in communal areas, including Colana, Gogela and Nozitshena, in departments of the state such as the Department of Agriculture, Department of Environmental Affairs, and Department of Rural Development and Land Reform (Kapfudzaruwa and Sowman 2009). Environmental offences have been shifted from traditional courts to civil courts. For instance, a number of community members had been arrested for harvesting *Ptaeroxylon obliquum* (*mthathi*) and hunting wild animals for food.

Although the South African constitution recognises traditional leadership as the institution that governs rural communities, the state does not provide have regulations and policies that clearly spell out what actually its roles are besides that it presides over traditional ceremonies and promotes traditional culture. The Traditional Leadership and Governance Framework Act (41) of 2003, the most pro-traditional leadership law, only encourages officials of state departments to create partnerships with traditional leaders when implementing projects in rural communities (Republic of South Africa 2003).

Respondents reported that government officials take advantage of this policy gap to side-line local traditional leaders in implementing resource management programmes. A case in point is the Working for Water (WfW) programme. This programme is commissioned by the Department of Environmental Affairs which it implements through municipalities to clear alien invasive plants, especially *Acacia mearnsii* (*Ublekwati wemnyama*) and *Acacia dealbata* (*Ublekwati webomvu*) in order to restore water flow in riparian zones (Blignaut et al. 2010; Büscher 2012;

---

homeland called Translkei with its own governance system. The state exterted minimum influence in the homeland and traditional leaders enjoyed a lot of power over community affairs (Mamdani 1996; Ntsebeza 2008).

Murata et al. 2019). Respondents argue that, although this programme is implemented in their communities, their traditional knowledge system and local traditional leadership are not integrated in its design and implementation.

In addition, the municipalisation[4] of rural South Africa, which started in 1995 and was consolidated in the 2000 municipal demarcation exercise in which rural villages were co-opted into the municipal governance system represented by elected ward councillors at the community level added another institutional blow to the territorial power of traditional leaders. The major source of institutional pluralism in this case is that the state does not have legislation and regulations that explain how the municipal system and the traditional leadership system should co-exist in rural South Africa. While traditional leaders claim rightful jurisdiction on the basis of custom, the constitution and a plethora of legislative instruments entrust municipal officials with the responsibility and power to drive village development, including management and rehabilitation of ecological resources. Ntsebeza (2008) argues that it is difficult to make traditional leaders part of a liberal democratic system because they are not elected. He argues that traditional leaders' practice of claiming legitimacy on heredity is despotic, hence it flies in the face of a democratic administrative system.

9.6.2.2 State Overreach in the Domestic Space

Respondents reported that they found it difficult to teach their children traditional management practices anymore because the democratic government interferes excessively in matters of household administration, especially in relations between parents and children. Generally, traditional management practices such as the ones that make it taboo to use certain resources (e.g. *Rapanea melanophloeos*) can only succeed if

there is a culture of strict obedience and respect for authority. Because they are young and don't have full understanding of TEK, children are generally likely to get tempted to disobey the taboo for quick gains such as getting fuelwood, drinking water or fruits. When children violate the management taboo, parents as the primary transmitters of TEK, may need to use punitive measures as means of enforcing obedience. However, the policies of the democratic government criminalise corporal punishment even when it is exercised by parents to their children within the domestic space. One elderly man at Gogela village said:

> Today's youth do not listen to us. When we discipline our children, government calls it abuse. We are living in democracy. You are even afraid of government, because you do not know what the government will say if you discipline your child. As parents start to lose control over their households, things fall apart. There is a lot of damage happening at household level because we no longer follow our customary laws. Instead, we are using the law of government. If I discipline my children today, social workers and police vans will be up and down these mountains now. So how do I teach them? When government said children must not be disciplined, that is the day things began to be ruined. We [parents] lost control over our families.

Unlike the facts and rules of scientific management methods whose cause-and-effect arguments which provide the compelling reasons why certain management techniques have to be implemented (e.g. that growth of *Acacia mearnsii* on riparian zones causes depletion of water resources due to excessive water loss through transpiration), how exactly use of *Rapanea melanophloeos* as fuelwood causes excessive mensuration of women cannot be easily explained to children. Children in particular and all other community members generally, do not need to understand the facts about some of the traditional management practices in order to implement or respect them. Instead they simply need to respect the truth-claims made in respect of these practices and obey the rules. This makes it necessary that there be strong relations of respect and obedience between children and parents.

---

[4] For details on the legislative position on municipalisation of rural South Africa and its distribution of power between municipal officials and traditional leaders refer to the White Paper on Local Government (Republic of South Africa 1998).

The role of parents and other community elders in transmitting TEK to children cannot be overemphasized (Berkes 2012; Cicin-Sain and Knecht 1995; Ohmagari and Berkes 1997). In a study of transmission of TEK among the Cree societies of Canada, Ohmagari and Berkes (1997) found that children are taught TEK by their parents through apprenticeship from a young age. At the study sites, parents use various strategies, including practical demonstration, storytelling and punishment to teach children. The major implication of this is that the state should respect the role of parents in household administration and stop its overreach tendencies.

Administration, as well as formulating rules and enforcing them, needs to be respected.

State overreach can be analysed using the realist dimensions of context and dynamism. The democratic state's policy of intervening in the domestic space and reconfiguring parents-children relations provides a contextual factor that bears significant implications to the processes of teaching traditional management practices. Respondents said that without the power to enforce discipline, it was difficult to teach traditional ecological knowledge both within the domestic space and at community level. They argued that some key aspects of TEK such as taboo, sacredness and respect for ancestral power, can only work when people are disciplined and obedient. Owing to the fact that these are matters of faith, traditional practices do not leave much room for questioning and therefore some strict disciplinary regimes must be enforced. This is a critical contextual factor that is lacking in the contemporary domestic space.

The state and the configuration of its relations with local communities and citizens are not static, but dynamic. The state has over the years evolved from being pre-colonial, colonial, apartheid and democratic. At every stage of the evolution, the state provides different opportunities and challenges to the role of parents as transmitters of TEK. For instance, the apartheid government was not concerned about civil rights of black people, and hence it did not interfere much in the local administrative arenas, especially the household space, of traditional communities. In contrast, in its pursuit to enforce the constitutional provisions of equality between persons, the democratic government actively intervenes in matters of household administration. This dimension of dynamism can help explain why in the past, even when the state system was already in place, traditional management practices fared better during colonialism which used indirect rule (Mamdani 1996) and apartheid which set aside the study region as a homeland (Ntsebeza 2008) than it does in the present democratic state.

9.6.3 Lifestyle Challenges Confronting Traditional Management Practices

Lifestyle challenges to TEK concern attitudes, tastes, values, preferences and outlooks that constitute the character of an individual or community's mode of life (Collins English Dictionary 2005; Oxford Advanced Learner's Dictionary 2015). Lifestyle is a key enabler or constraint to implementation of traditional management practices because, as scholars have noted, TEK is not just a way of knowing; it is practical in the sense that it is a way of doing things that individuals and communities need in order to support their wellbeing, including health, diet, spirituality and shelter (Berkes 2012; LaDuke 1994). The nature of support that traditional management practices may contribute to people's wellbeing at a particular point in history may exert significant influence on how the people perceive the relevance of these practices. The major lifestyle challenge that confronts traditional management practices at the three communities is integration of rural life into the market economic system.

9.6.3.1 Integration of Rural Life into the Market Economic System

The key lifestyle challenge facing TEK is that local communities have significantly shifted from a lifestyle which, in the past, heavily depended on direct support from ecosystem services to a market-based mode of living characterised by commodity accumulation. A substantial proportion of respondents (88%) reported that because of improved market linkages between the countryside and centres of industrial production, their fellow community members are able to carry on life with less dependence on nature's services. For instance, it is now possible for local people to build nice houses out of zinc and brick bought from the market, instead of going into the forest to cut timber and grass. Houses built of bricks and zinc often look smarter and are more durable than those built of timber and grass. This market development has significantly reconfigured people's perception of the value of nature to their construction needs.

Consequently, current generations look down upon those who build houses using natural resources as poor and backward. This emerging perception is also reinforced by the government's Reconstruction and Development Programme (RDP) which builds brick and zinc houses for the indigent citizens. The RDP stipulates that human shelter must be decent. The RDP's concept of a decent house is a shelter with brick walls, glass windows, zinc roof and ceilings (Republic of South Africa 2004). Nearly three-quarters of the respondents (74%) said these developments have rendered traditional management practices such as designating some forests resource as sacred in order to regulate harvesting of their timber products unnecessary because the timber is no longer as important as it was in the past. Of course, there are still some instances in which local people use traditional building material, especially when building cultural shelter such as rondavels and *amaboma* (temporary huts built in the forest for initiation of boys), but this does not result in high demand for natural resources.

Changes in lifestyle also adversely affect application of TEK in the management of ecosystem services that support livelihoods. Historically, subsistence agriculture in the form of dryland crop cultivation and livestock ranching was the key pillar of rural livelihoods for the communities of the Eastern Cape (Bundy 1988). Respondents spoke fondly of the past during which they lived on farming of which main crops were maize, pumpkins and beans, which collectively, constituted the local staple diet. Success in farming depended mainly on the implementation of appropriate traditional management techniques which included maintaining soil fertility, preventing erosion, planting at appropriate times of the agri-ecological calendar, and harvesting in-field water (Denison and Wotshela 2009). The impetus to implement these management strategies is undermined by widespread accumulation of money as a consequence of increased job opportunities found in major urban and mining centres, as well as the social grant system. This has led to monetization of rural livelihoods of which one major consequence has been a progressively growing trend of deagrarianization (Bundy 1988; Andrew and Fox 2004). Accumulation of money enables households to satisfy all their livelihood needs, including vegetables, legumes, meat and fruits, through buying from grocery shops.

In the past, when local communities obtained most of their food from farming, TEK was embraced because it provided the wisdom that they needed to support productive farming. Such wisdom taught households how to implement management strategies, including *gelesha* which involves turning the soil before planting, to harvest and store in-field water from rain, snow and dew in preparation for planting, stone terracing for preventing soil erosion, filling up dongas in order to rehabilitate degraded land parcels, and a taboo system for rainwater management. Previous studies at Gogela village, found that stone terracing supported the community over a long historical period, beginning from the late 1800s to the present (Denison and Wotshela 2009; Murata 2010). However, all these traditional practices are not taken seriously anymore because the livelihoods of the local people do not directly depend on nature anymore.

Furthermore, in the past, local communities heavily depended on medicinal plants for the health of their livestock, leading them to develop some management practices such as taboos used to protect trees from overharvesting. The taboos prohibited resource users from using some of the medicinal trees and shrubs for domestic purposes, especially fuelwood and construction. However, in the present management of medicinal plants is not taken seriously anymore largely because livestock farmers now depend on scientific methods such as dipping and immunisation for livestock health. The state made dipping compulsory starting from the colonial era in the 1900s and established trading stores of veterinary medicine in the local towns of Mt. Frere, Mt. Ayliff and Kokstad. This marked the beginning of the fall of ethno-veterinary practices among the communities (Beinart and Brown 2013).

Moreover, TEK was the key source of human health in the past, before the state made the scientific health system universally accessible at the inception of democracy. Traditional healers and some herb specialists were the main providers of health services, using traditional methods. Knowledge of medicinal plants, including their names, where they were found, whether in dry areas or wetlands, the time of the agro-ecological calendar during which they were found, and how to administer them to patients, was highly developed in the past. A sizeable number of respondents (40%) boasted that knowledge of traditional medicine was so strong in the past that they never went to clinics or consulted doctors when they were growing up.

Currently, knowledge of traditional health systems is vanishing as state officials run health campaigns in which they instruct local people to use clinics and hospitals for medical services.

Consequently, respondents attributed the loss of knowledge of traditional medicine to the rise of the scientific health care system. About 45% of respondents said they suspected that new[5] diseases such as HIV/AIDS, sugar diabetes, cancer and hypertension were brought on by scientific medical practices, especially immunisation injec-

tions that are administered to people during infancy.

The changes in lifestyle and their implication to TEK can be explained by critical realism's dimension of dynamism. Traditional management practices might have worked in the local communities in the past, but this is not a guarantee that they can work in the present, nor will in the future. Communities are dynamic; they continuously evolve and adapt to emerging conditions that bear an effect to their existence; needs and preferences. The conditions that might have been present in these communities in the past, may no longer be there in the present. This analytical argument is very important because traditional ecological knowledge is closely tied to the welfare needs and preferences of people and their communities (Berkes 2012) which, in turn, constitute the necessary conditions for involved individuals and communities to assess the value and relevance of any particular practice, including traditional resource management. This, in turn, is a decisive factor as to whether TEK is embraced or rejected by the same communities over different eras of their history.

9.7 Conclusions

This work makes three major contributions to the development of literature on the use of TEK in the discipline of resource management, including ecosystem services in the contemporary democratic South Africa. First, it demonstrates that efforts to apply TEK in contemporary South Africa are constrained by several challenges. This finding goes a long way to addressing the question why rural communities of South Africa's Eastern Cape Province are heavily degraded in spite of their rich historical wealth of traditional ecological knowledge. This is a big question that, ironically, is commonly overlooked by advocates and scholars of traditional ecological knowledge who mainly concern their work with arguing the potential value of the knowledge system. Furthermore, it presents an empirically informed rebuttal to the critique of traditional ecological knowledge offered by Du Toit, among others

---

[5] These diseases are fairly new in the epidemiological history of South Africa, and Africa broadly. HIV/AIDS started to be a commonly known disease in South Africa in the late 1990s. Even then, the country experienced widespread denialism shared among rural people and government technocrats until a decade into the millennium when he government adopted the antiretroviral policy. The government of Thabo Mbeki (1999–2008), including his Minister of Health Med Dr. Manto Msimanga-Shabala, denied that HIV causes AIDS: On the other hand, sugar diabetes and hypertension were perceived as rich people's diseases.

(Du Toit et al. 2004; Smith and Wishnie 2000). By and large, the critique argues that the failure of TEK to effectively solve contemporary environmental problems is endogenous to itself, in the sense that the knowledge system does not possess appropriate epistemic tools to confront the complexities of present-day environmental challenges. In contrast, this work has demonstrated that there is a horde of factors exogenous to TEK that constrain its ability to provide solutions to contemporary environmental challenges. Conditions in contemporary South Africa do not allow space for full implementation of TEK in resource management. Consequently, analysis of degradation of traditional landscapes including the one by Fabricius et al. (2006) and Palmer and Bennett (2013) should consider the reality that there is a host of factors that obstruct full-scale implementation of the traditional management practices.

Second, the work has unravelled the interdependence that exists between society and knowledge. The configuration of this relationship at any given era of a community's history lays the condition for either acceptance or rejection of the knowledge system. The work achieves this by employing critical realism's dimensions of context and dynamism as analytical tools. In addition, the work uses the concept of hermeneutics to explain how the practice of interpretation influences local people's perception of TEK. Together with the dimension of dynamism, hermeneutical analysis demonstrates that as communities evolve as part of the historical process, new worldviews emerge, which in turn change how people perceive reality. In light of this, advocates and scholars of TEK can potentially enrich their work, especially recommendations, if they factor in the reality that knowledge has evolved in the traditional communities.

The third contribution is that the work streamlines the challenges into structural categories which can go a long way towards helping decision makers design and implement sector-focused, as opposed to generic, intervention programmes to address the challenges. The work achieves this by employing abductive analysis, which makes connections between individual cases with wider structural mechanisms through moving research inquiry from the level of the empirical down to the level of the transfactual. This reveals how ideographic events and perceptions are in actual fact just the tip of the iceberg of a wider complex of structural forces that shape the functioning, including worldview, institutional arrangement and lifestyle of communities at any given point in the historical process. In the final analysis, this work argues that all the challenges that confront TEK in contemporary South Africa are a consequence of adverse incorporation of rural traditional systems into the nation-state (Du Toit et al. 2007). Thus, even though the "plug-in" method of knowledge integration holds best for the integration of TEK and introduced knowledge systems in South Africa, the challenges are systemic and formidable and needs broader sensitization and advocacy for people to go back and appreciate some TEK which the system convinced them to abandon. It is even more difficult, given the fact that the youth has almost no knowledge of most of the TEK.

References

Ahmed SJ, Stepp R, Toleno RAJ, Peters CM (2010) Increased market integration, value, and ecological knowledge of tea agro-forests in the Akha highlands of southwest China. Ecol Soc 15(4):27. http://www.ecologyandsociety.org/vol15/iss4/art27/

Andrew M, Fox RC (2004) 'Undercultivation' and intensification in the Transkei: a case study of historical changes in the use of arable land in Nompa. Shixini Dev South Afr 21(4):687–706

Ayaa DD, Waswa F (2016) Role of indigenous knowledge systems in the conservation of the bio-physical environment among the Teso community in Busia County-Kenya. Afr J Environ Sci Technol 10(12):467–475

Bank L (2002) Beyond red and school: gender, tradition and identity in the rural Eastern Cape. J South Afr Stud 28(3):631–649. https://doi.org/10.1080/0305707022000006558

Bank L, Southall R (1996) Traditional leaders in South Africa's new democracy. J Leg Plur Unoff Law 28(37–38):407–430

Battiste M, Youngblood J (2000) Protecting Indigenous knowledge and heritage: a global challenge. UBC Press

Beinart W (1989) Introduction: the politics of colonial conservation. J South Afr Stud 15(2):143–162

Beinart W (2003) The rise of conservation in South Africa: settlers, livestock and the environment, 1770–195. Oxford University Press, Oxford

Beinart W, Brown K (2013) African local knowledge and livestock health: diseases and treatment in South Africa. Wits University Press, Johannesburg

Bennett JE (2013) Institutions and governance of communal rangelands in South Africa. Afr J Range Forage Sci 30(1–2):77–83

Berkes F (2012) Sacred ecology, 2nd edn. Routledge, New York

Berkes F, Folke C (1998) Linking social and ecological systems for resilience and sustainability. In: Berkes F, Folke C (eds) Linking social and ecological systems: management practices and social mechanisms for building resilience. Cambridge University Press, Cambridge, pp 1–25

Bhaskar R (2016) Enlightened common sense: the philosophy of critical realism. Routledge, London and New York

Blignaut J, Mander M, Schulze R, Horan M, Dickens C, Pringle C, Mavundla K, Mahlangu I, Wilson A, McKenzie M, McKean S (2010) Restoring and managing natural capital towards fostering economic development: evidence from the Drakensberg, South Africa. Ecol Econ 69:1313–1323

Brundtland G (1987) Report of the World Commission on Environment and Development: our common future. United Nations General Assembly document A/42/427

Bundy C (1988) The rise and fall of the South African peasantry, 2nd edn. David Philip Publications, Cape Town

Büscher B (2012) Payments for ecosystem services as neoliberal conservation: (reinterpreting) evidence from the Maloti-Drakensberg South Africa. Conserv Soc 10(1):29–41

CBD (Convention on Biological Diversity) (2004) Addis Ababa principles and guidelines for the sustainable use of biodiversity. Secretariat Convention on Biological Diversity, Montreal

Chalmers N, Fabricius C (2007) Expert and generalist local knowledge about land cover change on South Africa's Wild Coast: can local ecological knowledge add value? Ecol Soc 12(1):10

Cicin-Sain B, Knecht RW (1995) Analysis of Earth Summit prescriptions on incorporating traditional knowledge in natural resource management. In: Hanna S, Munasinghe M (eds) Property rights and the environment. Beijer International Institute of Ecological Economics and the World Bank, Washington, DC

Collins English Dictionary (2005) HarperCollins Publishers, UK

Costanza R (2008) Ecosystem services: multiple classification systems are needed. Biol Conserv 141(2):350–352

Cousins B (2008) Contextualizing the controversies: dilemmas of communal tenure reform in post-apartheid South Africa. In: Claassens A, Cousin B (eds) Land power and custom: controversies generated by South Africa's Communal Land Rights Act. University of Cape Town Press, Cape Town, pp 3–31

Critchley WRS, Netshikovhela EM (1998) Land degradation in South Africa: conventional views, changing paradigms and a tradition of soil conservation. South Afr 15:449–469

Danermark B, Ekstrom M, Jakobsen L, Karlsson JC (2002) Explaining society: critical realism in the social sciences. Routledge, London

Denison J, Wotshela L (2009) Indigenous water harvesting and conservation practices: historical context, cases and implications. Water Research Commission Report No. TT 392/09. Water Research Commission, Pretoria

Du Toit JT, Walker BH, Campbell BM (2004) Conserving tropical nature: current challenges for ecologists. Trends in Ecology & Evolution 19(1):12–17

Du Toit A, Skuse A, Cousins T (2007) The political economy of social capital: chronic poverty, remoteness and gender in the rural Eastern Cape. Soc Identities 13(4):521–540

Fabricius C, Scholes R, Cundill G (2006) Mobilizing knowledge for integrated ecosystem assessments. In: Reid WV, Berkes F, Wilbanks T, Capistrano D (eds) Bridging scales and knowledge systems: concepts and applications in ecosystem assessment. Island Press, Washington, DC, pp 165–184

Gomez-Baggethun E, Reyes-Garcia V (2013) Reinterpreting change in traditional ecological knowledge. Hum Ecol 41(14):643–647

Gómez-Baggethun ERIK, Mingorria S, Reyes-García VICTORIA, Calvet L, Montes C (2010) Traditional ecological knowledge trends in the transition to a market economy: empirical study in the Doñana natural areas. Conserv Biol 24(3):721–729

Guest G (2002) Market integration and the distribution of ecological knowledge within an Ecuadorian Fishing Community. J Ecol Anthropol 6(1):38–49

Guthiga P, Newsham A (2011) Meteorologists meeting rainmakers: Indigenous knowledge and climate policy processes in Kenya. IDS Bull 42(3):104–109

Jackson S, Storrs M, Morrison J (2005) Recognition of Aboriginal rights, interests and values in river research and management: perspectives from northern Australia. Ecol Manag Restor 6(2):105–110

Kapfudzaruwa F, Sowman M (2009) Is there a role for traditional governance systems in South Africa's new water management regime? Water SA 35(5)

LaDuke W (1994) Traditional ecological knowledge and environmental futures. Colo J Int Environ Law Policy 5:127

Mamdani M (1996) Citizen and subject: contemporary Africa and the legacy of late colonialism. David Phillip

Mantzavinos C (2020) Hermeneutics. In: Zalta EN (ed) The Stanford encyclopedia of philosophy. https://plato.stanford.edu/archives/spr2020/entries/hermeneutics/

McCarter J, Gavin MC (2014) Local perceptions of changes in traditional ecological knowledge: a case study from Malekula Island, Vanuatu.

Ambio 43:288–296. https://doi.org/10.1007/s13280-013-0431-5

MEA (Millennium Ecosystem Assessment) (2005) Ecosystems and human well-being: a framework for assessment. Island Press, Washington

Meer T, Campbell G (2003) Traditional leadership in democratic South Africa. Democracy Development Programme, Durban

Merriam SB, Tisdell EJ (2016) Qualitative research: a guide to design and implementation, 4th edn. Jossey-Bass, San Francisco

Murata C (2010) Stone terracing for crop cultivation in Gogela settlement of the Transkei, Eastern Cape, South Africa. MPhil. thesis. University of Fort Hare, South Africa

Murata C (2020) Indigenous knowledge of ecosystem services in rural Eastern Cape, South Africa. PhD thesis. Rhodes University, South Africa

Murata C, Mantel S, De Wet C, Palmer AR (2019) Lay knowledge of ecosystem services in rural Eastern Cape Province, South Africa: implications for intervention project planning. Water Econ Policy 5(2):1–29

Murata C, Ndlovu L, Ganvani L, Odume ON (2022) Demystifying contemporary customary land tenure in legally plural southern Africa. J Law Soc Dev 9:1–17

Ngara R (2013) Indigenous knowledge systems and the conservation of natural resources in the Shangwe community in Gokwe District, Zimbabwe. Int J Asian Soc Sci 3(1):20–28

Nompozolo S (2000) An analysis of the characteristics and constraints of smallholder commercial farmers in the Transkei region, the Eastern Cape, South Africa. Master's dissertation. University of Fort Hare, South Africa

North D (1990) Institutions: institutional change and economic performance. Cambridge University Press, Cambridge

Ntsebeza L (2004) Democratic decentralisation and traditional authority: dilemmas of land administration in rural South Africa. Eur J Dev Res 16(1):71–89

Ntsebeza L (2008) Chiefs and the ANC in South Africa: the reconstruction of tradition? In: Claassens A, Cousins B (eds) Land power and custom: controversies generated by the South African Communal Land Act. University of Cape Town Press, Cape Town, pp 238–261

Ohmagari K, Berkes F (1997) Transmission of indigenous knowledge and bush skills among the Western James Bay Cree women of Subarctic Canada. Hum Ecol 25(2):197–222

Oxford Advanced Learner's Dictionary (2015) International student's edition, 9th edn. Oxford University Press, Oxford

Palmer A, Bennett J (2013) Degradation of communal rangelands in South Africa: towards an improved understanding to inform policy. Afr J Range For Sci 30:56–63

Pearce P, Ford J, Willox AC, Smit B (2015) Inuit traditional ecological knowledge (TEK), subsistence hunting and adaptation to climate change in the Canadian Arctic. Arctic 68(2):233–245

Pieroni A, Quave C, Santoro RF (2004) Folk pharmaceutical knowledge in the territory of the Dolomiti Lucane, inland southern Italy. J Ethnopharmacol 95:373–384

Pilgrim SE, Cullen LC, Smith DJ, Pretty J (2008) Traditional ecological knowledge is lost in wealthier economies and communities. Environ Sci Technol 42(4):1004–1009

Ray ID (1996) Divided sovereignty: traditional authority and the state in Ghana. J Leg Plur Unoff Law 37(38):181–202

Republic of South Africa (RSA) (1996) Constitution of the Republic of South Africa (18 of 1996). Government Printers, Pretoria

Republic of South Africa (RSA) (1998) White paper on local government. Department of Provincial and Local Government, Government Printers, Pretoria

Republic of South Africa (RSA) (2003) White paper on traditional leadership and governance. Department of Provincial and Local Government, Government Printers, Pretoria

Republic of South Africa (RSA) (2004) A comprehensive plan for the development of integrated sustainable human settlements. Department of Human Settlements, Government Printers, Pretoria

Reyes-Garcia V, Vadez V, Byron E, Apaza L, Leonard WR, Perez E, Wilkie D (2005) Market economy and the loss of folk knowledge of plant uses: estimates from the Tsimane' of the Bolivian Amazon. Curr Anthropol 46(4):651–656

Reyes-García V, Guèze M, Luz AC, Paneque-Gálvez J, Macía MJ, Orta-Martínez M, Pino J, Rubio-Campillo X (2013) Evidence of traditional knowledge loss among a contemporary indigenous society. Evol Hum Behav 34:249–257

Sayer A (2000) Realism and social science. Sage Publications, London

Scorer C, Mantel S, Palmer AR (2019) Do abandoned farmlands promote spread of invasive alien plants? Change detection analysis of black wattle in montane grasslands of the Eastern Cape. S Afr Geogr J 101:36–50. https://doi.org/10.1080/03736245.2018.1541018

Sillitoe P (2017) Indigenous knowledge and natural resource management: an introduction featuring wildlife. In: Sillitoe P (ed) Indigenous knowledge: enhancing its contribution to natural resources management. CABI Publishing, Oxfordshire, pp 1–14

Sklar R (1986) Democracy in Africa. In: Chabal P (ed) Political domination in Africa: reflections on the limits of power. Cambridge University Press, Cambridge, pp 17–29

Smith EA, Wishnie M (2000) Conservation and subsistence in small-scale societies. Rev Anthropol 29:493–524

Stacey M (1994) The power of lay knowledge: a personal view. In: Popay J, Williams G (eds) Researching the people's health. Routledge, London, pp 85–98

Tengo M, Johansson K, Rakotondrasoa F, Lundberg J, Andiamaherilala JA, Elmqvist T (2007) Taboos and forest governance: informal protection of hot spot dry forest in southern Madagascar. Ambio 36(8):683–691

Turner NJ, Turner K (2008) "Where our women used to get the food": cumulative effects and loss of ethnobotanical knowledge and practice; case study from Coastal British Columbia. Botany 86:103–115

Turner NJ, Ignace MB, Ignace R (2000) Traditional ecological knowledge and wisdom of aboriginal peoples in British Columbia. Ecol Appl 10(5):1275–1287

Von Benda-Beckman F (2002) Who's afraid of legal pluralism? J Leg Plur 47:1–47

Wynn D Jr, Williams CK (2012) Principles for conducting critical realist case study research in information systems. MIS Q 36:787–810

Blending Academic World with Community for Development: Plugging-In for Knowledge and Service

10

Saa Dittoh, Paul Kwame Nkegbe, and Anna Bon

Abstract

The educational system that was left behind in Ghana and other countries by colonial governments was largely a replacement, rather than an integrative, one. It was aimed at producing manpower to replace the colonial personnel that were leaving after independence. It was not aimed at addressing the needs and aspirations of local people, and it was urban and elite biased. The "revolutionary" government of J. J. Rawlings decided to right the wrongs in higher education by establishing a multi-campus University for Development Studies in 1992 in northern Ghana "to blend the academic world with the community". A practically oriented, community-based, and problem-solving pedagogy was instituted. This chapter explains the use of the "plug-in" concept in the implementation of the new approach, and analyses the achievements and challenges faced. Within a relatively short period, many graduates of the University can be found in most rural and several areas of the country working in government, private and NGO sectors. The community-based training has made them more marketable for work in rural areas and practically oriented endeavours compared to graduates of older universities.

Keywords

Community-based · Problem-solving · Third trimester field practical programme · Indigenous knowledge · Ghana

10.1 Introduction

The importance of linking knowledge with social inquiry instead of leaving them disconnected and isolated from action has been recognized in contemporary times, more than ever. Whereas, especially in developing countries such as Ghana, neighbourhoods, even adjacent to college and university campuses, struggle with a wide variety of challenges, university resources could become agents of skilled support. The reported societal and environmental challenges are multifaceted and include, for example, urban decay, poor sanitation, environmental degradation, growing economic inequalities, unmet basic needs of vulnerable children, families, and communities in the areas of education, healthcare, housing, criminal and juvenile justice and unemployment (Strand et al. 2003). Several authors including Boyer (1990) and Edwards and Marullo (1999)

S. Dittoh (✉) · P. K. Nkegbe
University for Development Studies, Tamale, Ghana

A. Bon
Vrije Universiteit Amsterdam,
Amsterdam, Noord-Holland, The Netherlands

© The Author(s) 2025
S. Dittoh et al. (eds.), *Integrating Indigenous and Scientific Knowledge for Sustainable Food Systems in Africa*, Sustainable Development Goals Series,
https://doi.org/10.1007/978-3-031-85512-2_10

have criticised the limited definition of scholarship as research in pursuit of 'new knowledge'. They suggest that other forms of scholarship could be considered—these are scholarship of integration, scholarship of application, and of pedagogy. In their view, these forms of scholarship were undervalued and neglected in terms of both faculty roles and institutional credibility in the universities. For example, the scholarship of application is needed to address the challenges faced by societies, hence, the universities and research institutions are to reconsider faculty reward systems and redirect their energy to develop the resources needed to solve societal problems (Strand et al. 2003). Invariably, the scholarship of integration pursues the betterment of existing ways, methods, or processes of dealing with the shortcomings of a society rather than replacing the natural solutions or what may be termed as the indigenous knowledge system.

10.2 Education as an Anachronistic Relic from Colonial Times

The educational system that was left behind in Ghana and other countries by colonial governments was largely a replacement, rather than an integrative one. It was aimed at producing manpower to replace the colonial personnel that were leaving after independence. It was not aimed at addressing the needs and aspirations of local people, and it was urban and elite biased. As such, in most parts of Africa, the educational systems, in the light of the societal needs and requirements, remained largely dysfunctional. The training of researchers has been so inadequate that there is generally no knowledge by academics on how rural community members view or perceive development and how they think of the problems of underdevelopment, deprivation, exclusion and others. Very few researchers in Africa are critical thinkers because of the defective educational system and its methodologies. There have been numerous attempts over time to reform and improve the educational systems, but successes have been minimal. Until present day, numerous high-level graduates from African universities, including engineers, economists and technicians are unemployed because there are hardly any industries needing their skills. Very few agriculturists from the universities have farms or are employed on farms because the agricultural training has not been geared to farming in Africa. Currently, agribusinesses are about selling inputs and agricultural products but not actual food production or processing. Who needs inputs if nobody is producing and how can there be products if there are no producers? There is so much disconnect between agricultural training and farming in Africa that agricultural development in most African countries continues to be a mirage.

The basic problem of the educational system is a lack of appropriation of the local African context in research and teaching methods. There is a lack of integration, of "plugging-in" into existing informal ways of teaching and learning, which are rather "apprenticeship-oriented" than purely theoretical. Another problem of conventional scientific theory-based research is that, in the lack of contextualization of knowledge—in which it differs from indigenous knowledge—it can hardly produce relevant results in the given context. While most problems, situations, activities in communities are systems-oriented, academic research has been "commodity-oriented". This has resulted in a disconnect between the needs of communities and the focus of academic research. Plugging-in into apprenticeship-oriented education is an alternative approach that has the potential to produce many more employable graduates. In this chapter, we focus on the scholarship of integration—to assess the significance of properly linking the university community (students, faculty, and their activities) to the developmental needs of local communities. By giving an account of the practically oriented, community-based, and problem-solving approaches of higher education (specifically, the University for Development Studies, Ghana) in community development, we argue that a university's interventions and mindset, when facing a supposedly negative societal challenge, must not be towards replacement, but towards integration.

The approach must be to analyse, help understand and appropriately fuse the existing knowledge with the new skills into an adaptive solution. This approach is what we refer to as the plug-in approach. It is an alternative to the idea of replacing indigenous knowledge and forcing externally invented technological solutions upon people.

According to Brown (2011), community-based research is a promising activity—collaborative in nature; change-oriented; and engaging to faculty members, students, and the community members in projects that address a community identified challenge. Involvement of the community or the supposed beneficiaries of a project or an intervention limits the potential failures of the intervention. In other words, the starting point for an effective intervention is the intended beneficiaries. Relevant stakeholders of the university and its various units play an important role in ensuring that the university impacts the society in terms of development (Jackson 2010).

The rest of the chapter is organised as follows. In the next section, education in northern Ghana dating back to colonial times is reviewed to place the chapter in proper context. This is followed by a brief on the University for Development Studies, and a discussion on indigenous knowledge and research at the UDS. The next section delves into the mission of the UDS and plugging-in for knowledge and service. The penultimate section presents some narrations of experiences of the third trimester field practical programme by former students of UDS, and that is followed by the concluding remarks section.

10.3 Education in Northern Ghana from Colonial Times

Formal education in several African countries has been traditionally elitist because of the history of education in those countries, the relatively high cost of education (compared to incomes) and the believe in many rural societies that they have to forgo the contribution of the labour of their children for many years. The development of formal education in northern Ghana from the colonial era till date typically reflects the above situation.

It is necessary to summarize the history of education in northern Ghana in this section to point out the euphoria that greeted the establishment of the University for Development Studies in northern Ghana in 1992, the important psychological impact it had on the people and the nation, and the expectations of the people in the area regarding its socio-economic and development impact.

The present area referred to as northern Ghana consists of about 90% of what was called the Northern Territories in the Gold Coast (now Ghana). The British colonial government administered the Northern Territories as a Protectorate. The first secondary school was established in the Northern Territories in 1951 in Tamale. At independence, in 1957, there were only two secondary schools (Tamale Secondary School and Saint Charles Minor Seminary/Secondary School in Tamale) serving the whole of northern Ghana, which covers 41% of the of the country and had about 25% of the population (Bening 2015). The colonial government did not see the need for many secondary schools in the Northern Territories because the aim of education according to the colonial government was to assist in governing the country, and minimum schooling was required. At one time, a Provincial Inspector of Schools in the Northern Territories, Lt. Col. M. F. G. Wentworth in 1937 said, "senior literacy education (i.e. middle school) will at the present stage be given only to able children who are required for specific Native Administration or Government posts of a clerical or technical type" (Wentworth 1937), and that "senior education was, and would for long remain selective: the 6 year Infant/Junior course was an entity in itself and would for a considerable period be the basic education of the country as a whole".[1] Also, the colonial authorities felt they had made some "mistakes" training people in the south and did not want to make the same mistakes with the people of the Northern Territories.[2] As stated by Der

---

[1] Minutes of a Meeting of the Northern Territories Board of Education, 18th February 1949. NRG 8/9/1, PRAAD, Tamale.

[2] Minutes of the Annual Conference of Officers in the Northern Territories, 21–24 December 1937. NRG 8/5/14, PRAAD, Tamale.

(1994), the colonial Governors in Accra viewed the Northern Territories as a drain on the resources of the Colony and Ashanti, and were therefore reluctant to spend money in developing it. It was clear that the colonial government had no interest in meeting the educational or developmental needs of the people of the Northern Territories. No wonder it (the colonial government) did not consider secondary school and by implication higher education necessary for people of the Northern Territories. It was even suggested in 1924 that all primary schools in the Northern Territories (except one) should be converted into junior trade schools as there were great prospects for artisans and those with practical training (Honter 1924).

After independence in 1957, the Government of Kwame Nkrumah took an opposite view from the colonial government and undertook massive educational development in northern Ghana (and indeed in all parts of Ghana) by providing many first and second cycle schools and education at all levels and it was largely free. That reduced considerably the cost of education for parents, but the issue of elitism and the perception of the people of loss of labour from their children persisted. Thus, northern Ghana continued to lag in terms of education. No tertiary educational institution was established in the northern sector until 1992, three and half decades after independence.

10.4 The University for Development Studies, Tamale, Ghana

The Provisional National Defence Council (PNDC) government of Jerry John Rawlings established the University for Development Studies by PNDC Law 279 as a multi-campus institution in the northern Savannahs of Ghana, to "blend the academic world with the community in order to provide constructive interaction between the two for the total development of northern Ghana, in particular, and the country as a whole". Thus, the University for Development Studies (UDS) by law, had to develop curricula that focused on community engagement and

development, and on poverty alleviation strategies, since its major area of operation was to be the poorest part of Ghana; the then Northern, Upper East and Upper West Regions. The Government had to be very emphatic about the direction of the new university because of general disappointments about higher education in Ghana and "new thinking in higher education which emphasized the need for universities to play a more active role in addressing problems of the society, particularly in the rural areas" (Effah 1998). There were national and international reservations about the wisdom of establishing a new University instead of using the resources to expand existing ones or establish research centres (Bening 2005), and the government of the day needed to justify its action. There were also concerns with the educational system in Ghana and several African countries generally. Several committees set up in Ghana to review the educational system that was inherited from the colonial days observed that there had been major shortcomings and deficiencies in the system (Ministry of Education 1972). The main concern was the need to restructure the educational system to ensure that school leavers will have employable skills. It was recognized that the educational curricula did not provide necessary attitudinal orientation and skills that could equip school leavers to be able to undertake practical work. It was, thus, generally argued that more technical and vocational institutions, rather than a university, were needed. The establishment of the university in the northern sector was, therefore, a significant milestone in educational development in Ghana, amid several misgivings and outright opposition. The UDS admitted its first batch of 40 students in September 1993 into the Faculty of Agriculture.

The first major challenge UDS authorities had was how to operationalize the University Law that makes it mandatory to "blend the academic world with the community". The first step that was taken was the institution of a trimester system where a third trimester of every year was to be devoted for engagement with communities. The modalities of engagement were however to be worked out over time, and thus between 1993 and 2002 several different methods were tried by the various Faculties. They included sending students on attachments to governmental, non-governmental and private sector establishments that operated in rural areas and sending students to communities to collect data. Though that helped students to have some experience of rural settings there was limited "blending" with the communities. It, however, helped the University authorities to think through and experiment various community engagement models. An innovative integrated Third Trimester Field Practical Programme (TTFPP) was finally arrived at, and it became operational from the 2002 academic year. The TTFPP became a university-wide programme in which all students at the University are expected to spend about 6–8 weeks (during a third trimester of every year) in communities. They are to stay in the communities for at least 6 weeks to interact, dialogue, plan and execute development initiatives with community members. Groups of about 10–15 students from diverse disciplines are sent to the communities. Staff members closely supervise the activities and students are scored on the field. They make oral presentations at the local government levels in which community members attend. They also submit written group reports for grading.

Apart from helping to meet some of the tenets of the Law establishing the university, the TTFPP system also afforded the University to overcome pertinent constraints. The university started on a small campus of an agricultural college with virtually no infrastructure; classrooms were very inadequate and there were hardly any laboratories. That situation, however, also provided an opportunity, and the university authorities decided to focus on the opportunity. It was argued that the "biggest and inexhaustible laboratories are the numerous communities across the country" (Dittoh 1997). What was required was innovation so that the various disciplines could find "laboratories" in the communities, and supplement that with whatever will be provided on the campuses. This unorthodox idea was strange at the beginning, but it was gradually accepted as realistic, innovative and should even be the "accepted" method of practical instructions for all disciplines. According to Dittoh (1997):

UDS's humble beginning is a mystery to many people. Though we lack many facilities, UDS students are acquiring experiences unknown in other universities. The experience of what to do in times of lack is a very good example. That experience and others are relevant and the future will prove that the problems we are going through will turn out to be our strengths.

The experience during the first couple of years of the third trimester system and other experiences at the local (community) level, especially through collaborative research and development work with local and international non-governmental organizations (NGOs), were what led to the idea of the Plug-In Principle.

10.5 Indigenous Knowledge and Research at the University for Development Studies

Either by luck, coincidence and/or divine intervention, collaborative research and outreach by several staff members of the UDS with national and international organizations, that believed in the central role of indigenous knowledge in development, started a few years after the commencment of academic activities at the university. One of the important collaborations was with the Information Centre for Low External Input Agriculture (ILEIA) of the Netherlands, and the Association of Church Development Projects (ACDEP) based in Tamale, Ghana. Staff of UDS and research stations based in Nyankpala in the Northern Region collaborated with these two organizations to form the Northern Ghana LEISA Working Group (NGLWG) with the main aim of promoting low external input and sustainable agriculture (LEISA) activities. As pointed out in Chap. 1, LEISA was one of the main vehicles for the promotion of IK and its integration with other knowledge systems. Researchers and development practitioners of the NGLWG showed from the output of the collaborative action research, the central role of IK in sustainable agricultural development and the usefulness of plugging-in to IK with appropriate conventional scientific knowledge (see Karbo et al.

1998; Alebikiya and Waters-Bayer 1999; Van Veluw 1999; Alebikiya and Karbo 1999; Dittoh 1999a; Dittoh and Alebikiya 1999; Kombiok et al. 1999; Dittoh 2001 and others).

Another collaborative research activity was with a GEF-funded People, Land and Environmental Change (PLEC) project which was led by the United Nations University (UNU), Tokyo, Japan. That project covered several countries in West Africa (Ghana and Guinea), East Africa (Kenya, Tanzania and Uganda), Southeast Asia (China and Thailand), Oceania (Papua New Guinea and Australia), North America (Mexico and Jamaica), and Latin America (Brazil and Peru). It was a participatory action research project which blended academic research and practitioners' knowledge (IK). The importance of IK in the project was very evident in all countries across the globe as documented in several book and paper publications (see Brookfield et al. 2003; Gyasi et al. 2004; Dittoh, 1999b; Anane-Sakyi and Dittoh 2001; Kranjac-Berisavljevic, 2003 and others). The emphasis of the project was restoration and maintenance of ecosystems, biodiversity and agrobiodiversity and it was necessary for the process to be guided by "people's science", that is, indigenous knowledge.

10.6   The University's Mission and the "Plug-In" for Knowledge and Service

The University for Development Studies adopted the baobab tree (*Adansonia digitata*) as its logo, representing resilience, and "knowledge for service". The University also undertook a strategic plan and came up with "the home of world-class pro-poor scholarship" as its vision statement. Its mission statement is, "to attain the vision by promoting equitable and sustainable socio-economic transformation of communities through practically oriented, community-based, problem-solving, gender-sensitive and interactive research, teaching, learning and outreach programmes". That made the university unique from all other Ghanaian, and indeed African Universities. The

vision and mission statements literally erased the "ivory tower" mentality right from the inception of the University. More importantly, the experiences convinced the pioneer staff and students of the University of the importance of plugging-in to local (indigenous) knowledge. It was observed that it would be impossible to undertake "knowledge for service" if indigenous knowledge is ignored or downplayed.

The general methodology of education in Africa has largely been the method of trying to "civilize the uncivilized" and the "know-all" bringing knowledge to the "know-nothing". That had been and is still the tragedy of education in most parts of Africa. Education was meant to first help the colonialists to have access to Africa's natural resources by training lawyers who were usually from royal houses. The trained African lawyers were used to acquire large tracts of land and mineral (particularly gold) concessions for the colonialists under arrangements which they also benefited. Thus, the methodology of education, just as in other "modernization" models, was replacement rather than integration. Whatever indigenous methods of education that existed in Ghana and indeed most parts of Africa had to be replaced rather than integrated with what was brought.

Figure 10.1 illustrates the "knowledge for service" plug-in orientation of UDS field practical training pedagogy. The first-year young inexperienced students first learn community-level knowledge in diverse areas: skills, local values, ethics, mannerisms as well as local level problem-solving skills during the first time they stay in the communities, and then subsequently design appropriate interventions to plug-in to improve situations in the communities. That process has proved to be very beneficial to the students and community members over the years. This basic practical training has been for all the students. However, the Medical School of the University also went on to adopt the problem-based learning (PBL) methodology which emphasizes preventive rather than curative medical practice, and critical self-enquiry by students. That resonates very well with indigenous medical practice, even though there has not been any effective integra-

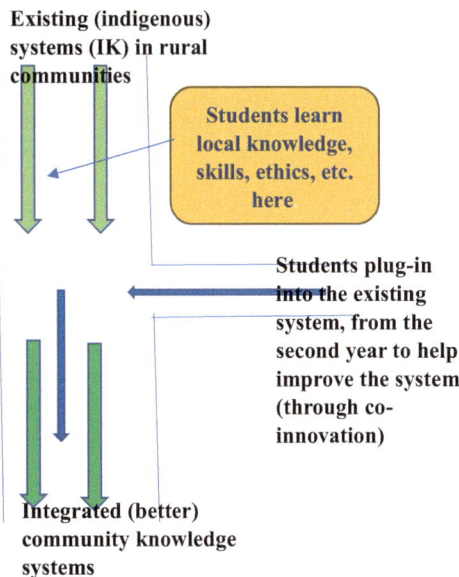

Existing (indigenous) systems (IK) in rural communities

Students learn local knowledge, skills, ethics, etc. here

Students plug-in into the existing system, from the second year to help improve the system (through co-innovation)

Integrated (better) community knowledge systems

The young inexperienced students must, first, plug-in with mindsets of people coming to learn from "illiterate" farmers and community members (hence the narrowness of the plug-in pipe in the diagram). They must learn, understand, and appreciate the prevailing system and practically experience the indigenous knowledge (IK) that exists. The understanding of the local knowledge (IK), skills, ethics etc. is important so that from the start they can connect IK and "modern science" that is being taught in the classrooms, and for the development of appropriate innovations and technologies that can be used to plug-in to integrate with the IK.

Fig. 10.1 "Knowledge for Service" Plug-in by UDS Students

tion of traditional and conventional medical practice in the UDS Medical School.

During a visit to the UDS in 2002 by the then United Nations Secretary-General, Mr. Kofi Annan, he said he was convinced that the TTFPP "approach should serve as an inspiration for all those who work for Africa's development, including especially for us in the United Nations" (Annan 2002).

10.7 Some TTFPP Experiences of Former Students of UDS

Below are the experiences of three former UDS students who went through the programme in different parts of the country and in different years:

1. I am Agatha Zeng.[3] I was an undergraduate student in the UDS between 2002 and 2006. I was among the first batch to start the integrated third trimester field practical pro-

---
[3]Madam Agatha Zeng is a Corporate and Investment Banker. She is currently the Head of Global Market Sales of an international bank.

gramme (TTFPP). Prior to our time, UDS students underwent attachments to governmental, non-governmental and private sector organizations to gain practical experience.

As I journey through the life that I've been blessed with, many are the things that I am grateful for, but few are the ones I consider **"absolutely invaluable"** and among those is the University for Development Studies' integrated Third Trimester Field Practical Programme (TTFPP).

I still clearly remember the horror and sheer panic I felt in my first year at the University for Development Studies, when it was announced to my batch of first year students that the model for the very much anticipated 3rd trimester programme had been revised and that our batch and those to come after us, would be sent to live in rural communities for 6 weeks in each academic year (*from first to third year*) in a bid to ensure we acquire a deeper understanding of the very society we were tirelessly studying to eventually contribute to their development. In our young minds, this was the most "inhumane" fate to be brought upon anyone. I mean, after toiling to secure the grades to enter this institution in hopes of being trained to enter the corporate world, we couldn't help but wonder what on earth the authorities were thinking. We protested with our all, but it eventually became obvious the minds of the University authorities were made up. Despite the widespread disappointment across the campuses among our peers, the dreaded day for us to be sent to our communities arrived and indeed, we were loaded on buses with our belongings and dropped off at our respective communities after long hours of travel on terrible roads.

To make matters worse, my group of 11 students (8 men and 3 women) were the last to be dropped off at our community, Klukpong in the West Gonja District of the Savannah Region (but it was part of the Northern Region then), and in the late evening when it was already dark. In fact, there were no other settlements after ours as we shared boundary with the Mole National Park. All we saw was total darkness as we were dropped off in a school compound with a community elder waiting to help us settle in. We were given two classrooms to reside in: one for the ladies and the other for the gentlemen. With our mattresses, mosquito nets, bags, etc. in hand, we started moving desks around in the classroom to enable us settle in and catch some sleep. I must say, if words could kill, our dear Professor Saa Dittoh, whom we all perceived as the brain behind our situation, wouldn't have lived to see the light of the next day, as I am certain several of our colleagues went through the exact chain of emotions as we did.

The next morning brought us to terms with the true reality of our situation. Our community had no electricity, no running taps, no health facility, no communication network, the only school in the community (*a six-classroom block*) had just one pupil teacher and a headmistress, plus a host of other shocking revelations. We had little kids all around and trying to peek at us through the windows. We later found out from their mothers that the narrative from the kids when they got home was that "some beautiful white ladies" were living in their classrooms. In fact, we were treated like "celebrities" with kids staring at us wherever we went.

As the days and weeks went by, our constant interactions with these amazing people in our community opened our eyes to the fact that we were truly privileged. Many were those who had never seen any other community aside theirs. The commitment with which the kids would show up in school each morning despite the extreme cold of the harmattan weather, and in tattered clothes, was inspiring and heartbreaking at the same time. With just two teachers for the entire school, these kids were clearly behind when compared to their peers. While we worked on our project and the related reports, some of us started complementing the efforts of the two teachers by teaching some of the classes in our spare time. It was such an eye opener just observing how grateful these kids were for the little we had to offer and the level of commitment they demonstrated as they did the best they could to make the best out of the little times they had us around.

Words could never describe the joy and sense of accomplishment that came with seeing some of these kids begin to grasp the concept of phonetics as they gradually mastered reading by themselves. We made some amazing friends among the women especially, who on countless occasions expressed their fascination about the fact that we as "girls" had progressed to the University level and could read and write just like our male colleagues. We made it a point to always encourage and assure them that their girls could do same if given the opportunity.

Our third year was the last time we stayed in this amazing community and although this is years back, I have since lived with this deep hunger and commitment to make a difference, no matter how small, wherever I find myself. How could I remain the ignorant and self-centred person I was before passing through this programme? My eyes were opened to just how privileged I have been and to the fact there's a lot more to life. This experience I must say has left a deep mark on me. Each day when I step into the shower to enjoy the limitless flow of water, I am reminded that there's a woman somewhere who must trek miles, just to have a basin of water for her household. When I drop my kids at school each morning, I remember there are kids under trees who do not even have the privilege of a desk nor a teacher to teach them, and that these kids will have to compete with my kids and others who are way more privileged than them for opportunities in their adult lives. When I think of all the amenities I have at my disposal, I remember the experience of falling sick in my community (Klukpong) and what it took to get me to the closest health facility. I constantly carry this burden upon my heart and in my small way, I always strive to make a difference wherever I find myself.

I look around me today, and can't help but wonder, the kind of leaders we would be churning out if each had this kind of exposure in their formative years. I will forever be grateful to UDS for this life-changing experience, and it is my wish and prayer that this amazing programme would be rolled out in other institutions as I strongly believe, the fibre of a person's being is dependent on what resides in his or her subconscious.

2. I am Charles Yaw Okyere,[4] a former student of the University for Development Studies (UDS) who participated in the University's Third Trimester Field Practical Programme (TTFPP) between 2003 and 2007.

The *"Third Trimester Field Practical Programme (TTFPP)"* presents fond memories during my bachelor's education in Agriculture Technology at the University for Development Studies (UDS), Nyankpala Campus. We were posted to Kusawgu, an important paramountcy in the Gonja Traditional Area. It is one of the five divisional skins (i.e., paramountcies) that ascends to the prestigious position as the Yagbonwura, the King of the Gonja Traditional Area. It is important to add that the late Yagbonwura Tuntumba Boresa I was a former Paramount Chief of the Kusawgu Traditional Area with the skin name Kusawguwura Sulemana Jakpa. I feel excited and honoured that the TTFPP made it possible

for me to have had some association with Gonja royalty.

We were hosted at the colonial administration block located in the community centre and close to the then chief's palace. This building was used as a police post after we left the community, although my recent visit in September 2022 shows that the police post was no more due to a few challenges. Interestingly, I noticed the extensive expansion of the community, compared to our stay some 15 years earlier.

We were well received by the indigenes of the community, and we enjoyed a lot of support from the chiefs, opinion leaders and community members. We had the privilege of fetching rainwater from the health centre's reservoir. I cannot forget the number of times cooked food and foodstuffs were brought to us. How can I forget the football games with community members? What about watching television in the house of "Liga Thunder" (the son of the late Kusawguwura)? I still remember the late Alhaji Popo (i.e., Pumpuni) who was our key informant in the community. I also enjoyed our teaching experience in the basic schools. We, within a short period of time, got integrated into the community.

It is important to note that field practical programmes as part of tertiary education in Ghana is not a new concept. For instance, the School of Agriculture (SoA) of the University of Ghana (UG) do send level 200 students for practical farm/field activities in the research institutes/centres for about 6 weeks during the long vacation. However, the TTFPP is unique in several aspects. First, the key feature of the TTFPP is its rural sociology and/or community development concept where students stay and integrate in the communities of diverse cultural backgrounds in northern Ghana. This offers a unique opportunity for the students to learn practical life lessons.

Second, the TTFPP has had its fair share of implementation challenges. Whilst we were very fortunate to have good accommodation and water by the community standard, it was not so for some of our colleagues in other communities. For instance, I can recall an experience where some of our colleagues were posted to communities with very few "zinc" roofed houses and they had

---

[4]Dr. Charles Yaw Okyere is a Senior Lecturer in the Department of Agricultural Economics and Agribusiness at the University of Ghana, Legon.

to stay in classrooms. Others had to live in unlocked/ungated rooms with occasional ransacking of their belongings. Other challenges included late departure to the various communities, security issues in few communities, late night assessment by supervisors using car headlights, poor road networks and limited access to mobile networks, among others. Upon proper reflection, who are we (i.e., the privileged students) to complain about the physical infrastructure? After all, this was the way of life for many people living in rural communities in several parts of Ghana. Therefore, I cannot overemphasize the importance of the TTFPP in the early exposure of students to living conditions in rural communities. These are practical experiences which have shaped my academic and research work.

Another important component of the TTFPP is its duration where students stay in the communities for three continuous academic years (i.e., Years 1–3) with each stay not being less than six (6) weeks. This is important in ensuring proper bonding among the team members and with community members. Since it is not a "one-off" event, this is expected to shape students' behaviour during their stay in the communities.

It is also important to highlight the multidisciplinary nature of the TTFPP (i.e., students with agriculture, medicine/health, humanities and natural science backgrounds forming one team to analyse challenges in the communities). This aspect helped the students to identify research problems from different disciplines/fields of study. This is an important element of providing students with the opportunity to collaborate at the early stage of their academic journey. I cannot forget how roles were distributed during the data collection, writing and presentation of the report. I can recollect an instant where our team members (Patricia and Cletus) were sent to Damongo (the then District capital) to collect weather and district socio-demographic information for our report, and they returned very late to the community because of the distance and the very bad road. Rehearsals for the presentation of the project report were fun. As an agricultural economist, the rural development component emanating

from the TTFPP distinguishes UDS' bachelor programmes from other universities in Ghana.

Let me conclude this short write up on my experiences from the TTFPP on how UDS can leverage this novelty to enhance its academic and research programmes. The TTFPP presents a unique opportunity of building a strong database for the local economic development of northern Ghana. For instance, the TTFPP could be used to collect large primary data to enhance community development of northern Ghana. Well established and accessible portals based on the TTFPP's data could be used to analyse the development needs of rural communities in northern Ghana. The rich data from the TTFPP should be disseminated during conferences/seminars and this could also form the basis for state and private institutions to design local economic development policies and programmes for northern Ghana. Therefore, the TTFPP could be an important database for tracking the development of rural communities in northern Ghana. However, this could be achieved by securing funds to establish and maintain the database and to design standard survey instruments to ensure uniform data collection across the rural communities in northern Ghana. I cannot conclude without saying a big thank you to UDS, community leaders and members, team members and lecturers for their support during my stay in Kusawgu during the TTFPP and the opportunity to obtain a life-changing experience.

3. My name is Patience Atitsogbey.[5] I entered UDS in 2005 and completed in 2009 with a BSc degree in Community Nutrition.

I undertook my third trimester field practical training programme (TTFPP) in Kwabia in the Kintampo South District of the Bono East Region, Ghana. Kwabia is a farming community and had a population of about 250 people at the time. It had a primary school and a Junior High School (JHS) that also served adjoining smaller communities. Kwabia had no electricity, no clinic, and only one borehole at the time. It was

---

[5]Dr. Patience Atitsogbey is a Tutor at the Accra School of Hygiene, Korle-Bu.

always a struggle to get to the community since the road from the main trunk road to it was very bad. It usually took about 30–45 minutes' drive from the nearest town, Kintampo.

We were a group of nine (9) students, seven gentlemen and two ladies from the then three UDS campuses (Tamale/Nyankpala, Navrongo and Wa) that were sent there. We were very disappointed at the first sight of the community; and deep within me, I regretted choosing to study in UDS. "What at all in God's name are we coming to do in a village without light and full of mud houses? How are we going to survive in this village?" were the questions we were asking one another. We were already informed during orientation sessions of the conditions we would face in the communities but what we saw was far worse than what we imagined. I felt I was not in any way prepared to stay in such a place for even a night. Our feelings and disappointments, however, did not change anything. The bus that sent us to the community left and we had to face realities. Our first consolation was that we met genuinely friendly community members who did not even seem to feel their deprivation.

With great courage, the men among us led the way, and we were taken to the Unit Committee Chairman's house by a friendly and cheerful teenager. After introductions, the Chairman informed us that our school authorities came earlier on to inform them of our coming, so they were expecting us. He sent message across and soon the whole community gathered. We were given a very warm welcome and the best accommodation in the whole community was given to us: a classroom, yes, because that was the only building constructed with cement blocks and aluminum roofing sheets; all the others were mud and thatch buildings.

The nine of us (men and women) were to share one classroom as our residence. That was when it dawned on us that we were informed earlier during our orientation that part of the UDS mission is to produce "problem-solving" professionals. We had to find a solution. We quickly devised a way, with our bed sheets and some ropes, we divided the room into two, a smaller place in one of the corners for the ladies while the

guys took the other sides. We laid our mattresses and fixed our mosquito nets. The urinals became our bathrooms, and for toilets, it was "free-range". This is something most of us had heard of but never knew we would practice it one day. By the evening, we were given a water tank, by the Unit Committee Chairman, who became our 'father', and teenagers filled the tank with water from the borehole. Foodstuffs such as yams, cassava, plantain, maize, and gari were supplied to us daily and occasionally accompanied with grasscutter meat. This, the community members did, with so much love, for about 6 weeks in each of the 3 years we had to be with them. The community easily became "our community" and we were then always eager to go to "our community" during the third trimester. We gladly welcomed this new, simple and impactful way of living. Their way of living—simple, joyful, helpful, brotherly/sisterly irrespective of ethnicity etc.— despite the level of deprivation, was something we appreciated very much. It impacted our lives and attitudes tremendously.

We also learnt practical agriculture by accompanying community members, both the elderly and youth, to their farms. We took part in planting and harvesting yams, cassava, maize and other crops, and asked many questions as was necessary since it was expected of us to learn much from the community so that we can produce a "community development proposal" for the (our) community.

By the third year, we became true "Kwabians". I learned gari processing from the women, gari processing was a common business in most of the households. The lessons learnt from Kwabia were numerous. We learnt teamwork, how to co-exist with anybody in peace, and how to solve problems. I have been impacted so much that I can stay under any condition and work without complaining. I learnt to find joy and happiness without the luxuries of life and how to show respect, love and help to the underprivileged.

We also did our best to impact the lives of the community members in several ways. We were teaching both the primary and the JHS pupils, we made new friends, and some became our lifetime friends. A teacher we met there gave us all the support; he would take our phones to Kintampo (the nearest place with electricity) to charge for us. We encouraged him to further his education and he is now a chartered accountant! Our stay there impacted the lives of the youth, some went back to school. Some teenage girls who dropped out of school because of pregnancy went back to school after they were encouraged by us to do so. I and my lady colleague became their role models. I feel very privileged to have studied in UDS and participated in the TTFPP. It was a life-changing experience indeed!

10.8 Concluding Remarks

Engagement of the university community (students and faculty) in rural community relationships gives them the opportunity to apply academic experiences in real world situations. It is widely known that students can apply their research methods skills, and other classroom activities effectively when they are confronted with real field issues. Therefore, blending the academic world with the community for development through knowledge sharing and service enhances the training of professionals who understand the needs of many rural communities and are also willing to live and work with rural communities as 'bettering' agents.

Effective integration of the university community and the rural communities builds trust among both parties, especially if the outcome of student practical learning activities including their research are well disseminated among key stakeholders of rural development. Invariably, the community-based research approach of the TTFPP of the UDS has the potential to enhance rural development policymaking if quality data about the development challenges of the community are collected by students. Indeed, this is well articulated by the former students who shared their experiences of the programme.

Unlike the traditional student research activities practised, which end as result publications, the community-based practical research approach of the UDS TTFPP mobilises the community to use findings to advocate for policy change, enhanced local resources as well as improved local practices (Kranjac-Berisavljevic and Abaidoo 2014; Mwingyine et al. 2017), exemplifying the "plug-in" for knowledge and service. Furthermore, the inter and trans-disciplinary nature of the programme combines various academic disciplines and specialised and indigenous knowledge of communities in the engagement, making it different from what goes on in the other public universities of Ghana and elsewhere. In other words, the TTFPP approach looks beyond replacing local knowledge systems with the so-called modern knowledge, instead, it attempts to fuse/integrate well, in a process known as plugging-in, to ensure sustainable community development in all aspects.

References

Alebikiya MA, Karbo N (1999) Sustainable agricultural development – the experience of the Northern Ghana LEISA Working Group. The Savanna Farmer 1(1):2–5

Alebikiya M, Waters-Bayer A (1999) (ILEIA) Northern Ghana Programme. LEISA (ILEIA Newsl Low Extern Input Sustain Agric) 15(1 & 2):36–37

Anane-Sakyi C, Dittoh S (2001) Agro-biodiversity conservation: preliminary work on in situ conservation and management of indigenous rice varieties in the interior savanna zone of Ghana. PLEC News Views 17:31–33

Annan K (2002) Remarks on receiving an honorary doctorate degree from the University for Development Studies. UDS Newsl 8(3):13

Bening RB (2005) University for Development Studies in the history of higher education in Ghana. Hish Tahawah Publication, Accra

Bening RB (2015) The History of Education in Northern Ghana. GAVOSS Education Plc. Accra, Ghana

Boyer EL (1990) Scholarship reconsidered: priorities of the professoriate. The Carnegie Foundation for the Advancement of Teaching, New York. https://www.umces.edu/sites/default/files/al/pdfs/BoyerScholarshipReconsidered.pdf

Brookfield H, Parsons H, Broohfield M (eds) (2003) Agrodiversity: learning from farmers across the world. United Nations University/UNEP/GEF, Tokyo

Brown KT (2011) A pedagogy of blending theory with community-based research. Int J Teach Learn High Educ 23(1):119–127

Der BG (1994) The development of education in Northern Ghana during the colonial era. J Inst Educ 3(1):100–115

Dittoh S (1997) Address to UDS staff and students on the TTFPP programme. Nyankpala Campus, Tamale, Ghana, 8 Mar 1997

Dittoh S (1999a) Sustainable soil fertility management: lessons from action research. LEISA (ILEIA Newsl Low Extern Input Sustain Agric) 15(1 & 2):51–52

Dittoh S (1999b) Participatory technology development approaches to PLEC activities in Northern Ghana. Paper presented at the 4th WAPLEC regional workshop on the theme "Participatory approaches to eco-development and agro-biodiversity conservation", Pita, Republic of Guinea, Sept 1999

Dittoh S (2001) From LEIA to LEISA farming systems in Northern Ghana: possibilities and challenges. The Savanna Farmer 2(1):21–25

Dittoh S, Alebikiya M (1999) Internalising PTD and LEISA in agricultural training. LEISA (ILEIA Newsl Low Extern Input Sustain Agric) 15(1 & 2):52

Effah P (1998) An Address at a Students' Symposium. 14th May, Nyankpala Campus, UDS, Tamale.

Edwards B, Marullo S (1999) Universities in troubled times-institutional responses. Am Behav Sci 42(5):754–765

Gyasi EA, Kranjac-Berisavljecvic G, Blay ET, Oduro W (eds) (2004) Managing agrodiversity the traditional way: lessons from West Africa in sustainable use of biodiversity and related natural resources. United Nations University Press, Tokyo

Honter RE (1924) Memorandum on northern territories primary education generally. ADM 56/1/88. PRAAD, Accra

Jackson ET (2010) University capital, community engagement, and continuing education: blending professional development and social change. Can J Univ Contin Educ 36(2):1–13

Karbo N, Bruce J, Otchere EO (1998) The role of livestock in soil fertility management. Technical report. ILEIA/NGLWG Collaborative Research, Tamale

Kombiok JM, Aalangdong OI, Salifu AZ (1999) Case study on non-burning and organic farming practices in Northern Ghana. Technical report. ILEIA/NGLWG Collaborative Research, Tamale

Kranjac-Berisavljevic G, Abaidoo CR (2014) Evolving approaches to university education in Africa: recent examples from Ghana. Afr J Sustain Dev 4(3):39–50

Kranjac-Berisavljevic G (2003) Enthusiasm for PLEC continues in northern Ghana. PLEC News and Views, New Series No. 3 pp. 21–23

Ministry of Education (1972) Report of the Education Advisory Committee on the proposed new structure and content of education for Ghana. 23 Oct 1972

Mwingyine DT, Aabeyir R, Fielmua N (2017) Linking academia and community: evidence from student-community engagement in Ghana. Ghana J Dev Stud 14(1):208–230

Strand KJ, Cutforth N, Marullo S, Donohue P (2003) Community-based research and higher education: principles and practices. John Wiley & Sons, San Francisco

Van Veluw K (1999) The Muteensa story as an example of the participatory technology development (PTD) process. The Savanna Farmer 1(1):6–8

Wentworth MFG (1937) Education and native administration, their interrelationship. NRG 8/5/14. PRAAD, Tamale, p 8

Rural Development and the ICT4D Plug-In Principle for Information and Communication Technologies

Francis Saa-Dittoh

Abstract

It is widely believed that modern technologies such as Information and Communication Technologies (ICTs) for Development (ICT4D) provide important innovative intervention instruments to help achieve food security in Africa and elsewhere in the Global South. This idea is a key tenet of development policies promoted for several decades by Global North/West governments, but one also finds it widely reflected in policy circles in the Global South. In this Chapter, we take a critical look at the potential, promises and actual results of advanced technologies and their effects on agricultural value chains and rural development. We do so by examining failures and successes of a range of recent ICT4D projects in which the present author has been involved. We contrast failures and successes in ICT4D projects so as to identify the main differences that characterize those that succeed versus those that fail. While in some cases success seems coincidental or by chance, there are highly persistent underlying reasons that distinguish failure from success. Accordingly, based on long-term field experiences, I formulate what I call the ICT4D Plug-in Principle. Its two keypoints are: (i) collaborative: advanced technology needs and interests are to be formulated from the ground up by local rural communities, mediated properly by ICT designers by using appropriate participatory action research methodologies; (ii) contextual: the "innovative" interventions foreseen by advanced ICT technologies must fit (or "plug-in") appropriately to existing human information and communication patterns and habits as exhibited by local communities, and it is an ICT designer job to secure this. I discuss both the more-than-established ICT technologies such as radio as well as consider the now fashionable Artificial Intelligence (AI) technologies for their potential and usefulness in the Global South. Deliberately following our formulated ICT4D Plug-In Principle is a way to ensure that ICT projects, whether traditional or advanced, especially those that target improvements in food security and rural livelihoods, are truly for development (4D).

F. Saa-Dittoh (✉)
University for Development Studies, Tamale, Ghana

Vrije Universiteit Amsterdam,
Amsterdam, Netherlands

Keywords

Food security · ICT · ICT4D · Voice-based · Agricultural value chains · Middlemen · Rural northern Ghana

11.1 Introduction

Information and Communication Technology (ICT) is a transformative force seen across sectors around the world today. Its impact on agriculture and rural development is increasingly being felt and, I would argue, increasingly being understood in its real import. We need to remember that ICT is a broadly defined set of technologies that not only make communication happen but also make the exchange of information happen. These are things like mobile phones, personal computers, the Internet, and radio, which, despite not being seen as a cutting-edge tech tool, is probably still the most impactful ICT in rural and agricultural development.

For development professionals and research experts concentrated on agriculture and food security, the relevance of Information and Communication Technology (ICT) is hard to overstate. Agricultural practices, especially those in rural and low-resource settings, have often been constrained by factors such as poor infrastructure and limited access to the kind of information that drives timely decision-making. ICT, with its promise of almost ubiquitous access to information, seems well positioned to rectify these situations across a number of relevant axes and to beam necessary information directly to farmers.

When we discuss Information and Communication Technology for Development (ICT4D), we are talking about the poised use of ICT to bring about economic, social, and cultural development (think SDGs), especially in regions that are at a socioeconomic disadvantage. This is a strategic use of ICT, which has historically occurred in the privileged world, not just for the purposes of communication but for fostering growth — the kind of growth that the World Bank calls "equitable" and, in the ICT4D movement, has been termed "inclusive." So, for rural development, ICT4D can potentially play a very large role in two or three key kinds of services: improving the agricultural value chain, helping farmers to increase their productivity; connecting farmers with markets; and providing essential services to communities that are severely underserved.

11.2 Sustainable Development Goals, ICT and the "Digital Divide"

Remarkably, the United Nations' Sustainable Development Goals as launched in 2015 hardly mention digital technology at all. Only SDG-9.c considers their role, stating that access to the Internet should be available to and affordable for everyone on the planet.

Since then, the positive potential of ICT in achieving the Sustainable Development Goals (as sketched above in the Introduction) has gained increasing awareness and recognition. For example, in 2022, in the wake of the Covid pandemic, in an alarming report about the Sustainable Development Goals,[1] the United Nations have made an appeal to the international community for urgent concerted action, showing that the international agenda for Sustainable Development for 2030 was getting seriously off-track. Among the mitigation strategies proposed to address these global challenges, digital technologies are seen as playing an important role. For low- and middle-income countries, access to the Internet, ICT readiness and national data ecosystems are deemed to be of vital importance for monitoring and policy making.

11.2.1 The Infrastructure View on the Digital Divide

The SDG account above draws attention to the fact that Internet penetration is highly uneven in the world, with large parts having no Internet access at all. The disparities are especially big between the Global North and the Global South. This is commonly referred to as the Digital Divide.

This widely held view makes the Digital Divide into predominantly a matter of lacking technological infrastructure, with policy and development solutions focused on bringing more

[1] See: https://desapublications.un.org/publications/sustainable-development-goals-report-2022 (Accessed 29 September 2024)

advanced digital infrastructural technologies to the Global South.

Indeed, the digital infrastructure issue is an important one (still today, although things are changing as well).

The infrastructure view on the Digital Divide generally sits well with traditional narratives about International Development. These stem from the late 1950's (see US Presidential Security Advisor Walt Rostow's The Stages of Economic Growth, 1960), to today's policies of, for example, USAID, the US Government (Foreign Affairs) Agency for International Development. The underlying idea (or ideology) is that economic growth is achieved in a linear way by introduction and transfer of technology from advanced countries to (more backward) developing countries.

These days, it is especially Artificial Intelligence (AI) that is popular in this technological solutionism of the West (see the many funding calls on the websites of USAID and other western development agencies). The earlier cited 2022 UN Report on SDGs is basically in line with this; it sees the use of digital technologies as an important way to solve outstanding SDG challenges.

11.2.2 Accounting for the Real Complexities of the Digital Divide

The *lack-of-infrastructure* issues in the Global South are undoubtedly crucial in understanding and combating the Digital Divide. But it would be an oversimplification to think that these are the only issues to address. Generally, the Digital Divide encompasses much more than technological infrastructure alone and is much more multifaceted and complex.

In my view, the Digital Divide is to be seen as a rather (over)simplified metaphor for the tensions, contradictions and political and social struggles regarding the Digital Transformations that are currently taking place across the world. If we want to overcome the Digital Divide, we have to account for a range of outstanding issues,

much richer and much more complex than infrastructure alone. Here are some corresponding take-aways[2] from recent digital technology and society international research that go beyond pure Internet access infrastructure issues:

[Literacy.] Internet and World Wide Web generally have a very text-oriented character. Even if access infrastructure is available and affordable, this excludes low-literate people, a serious problem in many countries, especially (but not only) in the Global South.

[Modality.] Even where people can read and write, speech is often a preferred modality of communication. For example, in many places people use WhatsApp to send voice messages rather than writing text.

[Language.] Internet and Web are dominated by content written in a relatively small number of big languages, such as English or Chinese. However, there are thousands of languages in the world, with billions of people speaking only their mother tongue. Support for many languages is inadequate or even totally lacking. Recent research shows that AI has potential here, however not by way of the extremely data and energy hungry Large Language Models (LLMs). AI methods are called for that can successfully employ small data corpuses and limited resources.

[Relevant content.] The Digital Divide is also widened because there is not much Web content relevant to the information needs of, say, the average smallholder farmer in the African Sahel. Commercial content producers and providers often do not see an interest in poor people.

[Indigenous knowledge.] Information and communication, to be effective, has to fit into the actual real-world contexts, world views, knowledge and practices of the people it is intended for. Although this is an old truism in innovation theory (Rogers 2003), it is com-

---

[2]For more evidence, cases and explanation, see the author's forthcoming PhD thesis "*From Radio to AI — African Community-Driven Development of Sustainable Information Systems*" (Amsterdam, 2025)

monly ignored in International Development projects and programmes (witness the Millenium Villages and OLPC projects of the early 2000's). Examples also abound in major sectors such as agriculture, education and health, and I provide some examples and insights below.

[**Education.**] In addition to the ability to read and write, there is also a need for what is called digital literacy, i.e. being able to use digital technologies appropriately and usefully. This is often seen as restricted to the acquisition of digital technical skills (such as searching the Web or making a Facebook page). But in view of recent developments such as the manufacturing of disinformation or AI deepfakes (or more generally, the political weaponization of digital technologies and social media), this must be broadened. Global citizenship requires more than purely technical skills, namely, the ability of critical thinking, weighing information, argument and their sources to come to conclusions about what is valid or valuable and what not.

These are all important factors that significantly influence and complicate the global landscape of the Digital Divide, and what to do about it. A binary distinction between (groups of) countries, such as Global North versus Global South, is inadequate to describe current realities and is misleading in designing policies for their betterment. An additional factor in this is what may be called the *urban vs. rural* divide. Within countries (also in the Global North) there are big discrepancies in development, both in terms of infrastructures and skilled human resources, between urban and rural regions. Rather than a binary divide, these uneven developments are better described in terms of collaborating as well as competing overlapping social networks, accompanied by different overlay networks also involved in political, social and power struggles (cf. Latham and Sassen 2005), and this simultaneously at local, regional and global levels.

Be that as it may, theoretically, a too narrow infrastructure view on the Digital Divide will *practically* trigger (unnecessary, regrettable and avoidable) failures of ICT4D projects, as I discuss below. The central point is that ICT4D projects, to be successful, have to take into account the full spectrum of Digital Divide issues as outlined above, and do so *in-the-field* and *on-the-ground*.

11.3 Why ICT4D Projects Fail

11.3.1 Disruption

There can be (and mostly is) the issue of disruption in the bid to introduce technologies to communities. Disruptions can cause ripples through society; disrupt social interactions and relationships, organizational structures, institutions, and public policies. Because innovation ecosystems are complex, dynamic, adaptive systems, technological design and development will by all means interact with social trends (Schuelke-Leech 2018).

An example of possible (theoretical) disruption came from my own research in voice-based marketing for agricultural products: a case study in rural Northern Ghana which led to the development of a voice-based prototype that allows medium to large-scale farmers in rural areas to place advertisements on the World Wide Web (WWW) (Dittoh et al. 2013). At the time, this seemed like a great idea, until further research showed that this solution would cut out "middlemen". These "middlemen" are community members who traditionally would liaise between farmers and markets and/or market to market. They may buy from farmers directly and sell at community markets, thereby incurring the cost and logistics of transportation, or they may also buy from one market and resell at another. The agricultural value chain will therefore have a number of these "middlemen" along the way from farmer to end-consumer. Along value chains, transportation naturally increases the prices, but also the multiple wholesaler and retail profits as well as several intermediaries with unclear functions further increase the final costs (Tracey-White 2005). The overall result is usually unremunerative prices for the agricultural products for farmers in rural areas, and high prices for consumers in urban areas. The persistence of such marketing arrangements results in chronic poverty and food insecurity for both rural and urban dwellers.

The prototype developed allows farmers to call directly into the system with a normal GSM mobile phone, get prompts in their own language which guide them to leave information on what produce they are selling (typically grains; maize, millet, rice, soybeans, etc), in what quantities (normally, in bags) and how much. This information is automatically formatted and placed visually on a website for prospective buyers. Technically, this system works well to this point until we suddenly find out that there is a problem; prospective buyers are unsure of the quality of the produce and would not fully trust that information from the seller. This scenario showed that although cutting out these middlemen with our digital ICT solution would possibly increase revenue for the farmer and/or lower the cost to the buyer, it was indeed a disruption of an already existing system; the quality of the produce, which hitherto I had brainstormed on how to handle, is determined and communicated by these very people I wished to cut out of the system. We will see later how we can circumvent this problem by way of the Plug-In Principle.

11.3.2 Non-acceptance

Apart from disruption, there is also the issue of non-acceptance of technologies. This is often expressed in a number of ways ranging from outright dismissal to accepting the technology and then gradually discarding it. An example from the early days of ICT interventions for developing countries were ICT kiosks; the creation of a set of touch-screen kiosks for remote rural communities in South Africa's North-West Province were initially well received by the communities. However, the kiosks' lack of updated or local content and lack of interactivity led to disuse, and the kiosks were removed less than one year later (Benjamin 1999). Similar setups were attempted in numerous places around the world, and simi-

larly, except in a few cases, they were gradually neglected. In India, by 2005, over 26,000 ICT Kiosks were set up in rural areas which was expected to serve 5.85 million rural farmers. Interestingly, the majority (almost 60%) of these kiosks were by the private corporate sector. Each kiosk costs roughly 66,000 Indian Rupees (over $800). Now, it was found that 92% of farmers were aware of the kiosks in their villages, but only 34% actually used them and all of these users were highly educated; tertiary (Sharma and Rao 2000). These two examples give the indication that something (or somethings) was missing in these setups; the latter case points to a problem of literacy.

Other examples from Ghana; the Accounts and Personnel Computerisation Project of Ghana's Volta River Authority (VRA) saw managerial staff pleased with the changes brought by the new system but apparently the system had 'bred a feeling of resentment, bitterness and alienation' amongst lower-level staff which led to resistance and non-use, particularly amongst older workers (Tettey 2000).

And finally, another personal example; sometime in 2008/2009, during the early days of the School of Medicine and Health Sciences of the University for Development Studies in Tamale, Ghana, an institute in the Netherlands (*name withheld*) took it upon themselves to provide the University with an ICT Kiosk. With great cost, the kiosk, which was quite large and equipped with state-of-the-art computers and accessories (but built to function only within the kiosk), was shipped from the Netherlands to Ghana's sea port in Tema, then transported over 500 kilometers by road to Tamale. A platform was built at the Tamale Campus of the University where the device was placed. After the initial opening and testing of the equipment, the kiosk stood unused and the internal equipment (now defunct and outdated) were dismantled and removed in 2021 (twelve years later!). That was clearly an ICT project failure and it was mainly due to non-acceptance. It began with the lack of proper context, and the result of making assumptions not founded on specific context. A representative of the institute in the Netherlands had been to Ghana and had in fact set up a similar kiosk somewhere in the North of Ghana; in a rural community. Upon hearing of a University that existed also in the North, and perhaps hearing of the community-based work the University students undertook, it was assumed that the campus was located in a rural area and this kiosk would be providing internet access, transmitted to it via point-to-point radio satellite dishes (which were provided); the representative had not (and never did) visit the University. There were some suggestions from the ICT staff at the University to consider alternative equipment, but the project was fixed and would only provide what they offered as is. The result; the University *did* have internet access already (using VSAT at the time); the students *did* go to communities, but to numerous communities and therefore one could not be selected for this purpose to the detriment of the others, and the equipment within the kiosk was set up in such a way that it was difficult to reuse for any other purpose. This led to the kiosk, which was meant to promote rural development, being abandoned for over a decade right in the middle of the university campus in an urban area.

11.4   Successful ICT4D Projects

11.4.1 Tibaŋsim

Tibaŋsim, one of the biggest successes of the author's PhD work, is not being touted as a success because it is currently in use but on the basis that it is currently sought after by all stakeholders that were involved in it and the rural communities in which it was piloted scored it an 80.50 on the System Usability Scale (Brooke 1996). Currently, it is an integral part of a $100,000 project funded by the Internet Society Foundation (TIBaLLi),[3] and is considered by the Internet Society to be the most innovative method thus far for information dissemination in low-resource environments and under review for funding to scale up to at least 200 communities. It was considered a great suc-

cess by community members and others mainly because of its contribution to improved efficiency in the agricultural marketing, and reductions in poverty and food insecurity in the targeted rural communities.

Tibaŋsim is an information delivery system designed for low-resource areas with limited access to digital sources. The system relies on GSM and FM radio in the local language(s) of the community. It was piloted in five (5) rural communities in the Savannah Region of Ghana, providing farming-related information to almost 1000 people.

The system is designed to bridge the digital divide, which remains a significant problem, with 4.54 billion people lacking access to information and communication technologies. The paper that describes Tibaŋsim highlights the importance of user-centered design and catering to contextual issues when developing such systems (Dittoh et al. 2021).

11.4.2 Foroba Blon (Voice-Based Citizen Journalism)

Foroba Blon, developed by the Web Alliance for Regreening in Africa (W4RA), is another successful example. Foroba Blon, which was deployed in Mali in collaboration with the Web Foundation, the VU University Amsterdam, and Sahel Eco, allows community members to call radio stations and leave audio messages from basic mobile phones to a web interface about weddings and lost cattle without being connected to the internet. Foroba Blon quickly became a local journalism tool and was winner of the Press Innovation Contest in 2011, sponsored by Google and issued by the International Press Institute,[4] where it was selected out of 376 submitted projects to support citizen journalism. It was also featured in the Los Angeles Times.[5] Foroba Blon drastically reduced the cost information dissemi-

---

[3] https://www.isocfoundation.org/2023/02/announcing-our-2022-research-grantee-cohort/

[4] https://ipi.media/ipi-news-innovation-contest-2011-delivers-3-winners/

[5] https://www.latimes.com/world/africa/la-fg-mali-social-network-for-illiterates-20190617-story.html

nation generally, and livestock marketing in particular.

11.4.3 Kasadaka

Both Tibaŋsim and Foroba Blon are built on a platform called the Kasadaka. The Kasadaka represents another perfect example of a successful ICT4D project. Kasadaka is a platform that supports easy creation of local-content and voice-based information services, targeted at currently 'unconnected'. Kasadaka means "talking box" in several Ghanaian languages (Baart et al. 2019). The platform consists of a combination of software and hardware, that allows for the hosting and development of innovative voice-based information services in the conditions of the developing world. The Kasadaka platform supports the formation of an ecosystem of decentralized voice-based information services that serve local populations and communities, built by local innovators and entrepreneurs. This is very much analogous to the services and functionalities offered by the Web, but in regions where Internet and Web are absent and will continue to be for the foreseeable future. The Kasadaka, as seen from the aforementioned projects has been the technical backbone of several successful ICT4D implementations and is also actively used for teaching students in a joint ICT4D in the field course[6] by the Vrije Universiteit Amsterdam, University of Malaysia (UNIMAS) and the University for Development Studies (UDS), Ghana.

11.4.4 The Mobile Phone

Finally, there is need to mention the biggest ICT4D success in history to date; the mobile phone!! Where other technologies, like fixed phone lines, dial-up internet, ICT kiosks, PDAs, broadband internet and the like never took root in Africa and other developing nations in the Global South, the mobile phone tells a different story.

This unexpected phenomenon has made waves everywhere in Africa and from shantytowns to remote villages, mobile phones are being used to market agricultural produce, transfer money, monitor elections, and deliver public health messages. A large informal economy has also emerged to support the mobile sector (Etzo and Collender 2010). Perhaps every successful African ICT implementation is built either directly or indirectly around the mobile phone. Kasadaka, Tibaŋsim, Foroba Blon and TIBaLLi are all built around the successes of the mobile phone.

11.5 Success Versus Failure

We come down to the vital questions; why did so many ICT for Development projects experience partial or total failures? And what led to the successes of some?

11.5.1 Stakeholder Involvement

The lack of proper stakeholder involvement will most likely lead to failure. Let's take a look at what this means; ICT projects are meant to solve problems, but in the case of developing countries and especially low-resource communities, one question we must ask ourselves is; who defines these problems? There is the tendency to perceive a problem that is not relevant within the local context. For example, the lack of access to the latest music album might be a relevant problem in certain places and totally irrelevant in others. Stakeholder involvement will give us an indication of what the real problems are and may even go as far as to provide the solution (in a non-technical form). This is however the foundational step, and requires other building blocks to ensure successful adaptation of technologies.

The main stakeholder of any ICT system is the end-user, which in most ICT4D projects, would be the members of a rural community. Apart from the community members, we also would normally have local industry or providers; e.g. farmer groups, food processing groups, local

---

[6]https://ict4dinthefield.com/courses/course-v1:VU+XM_0008+2021_p6/about

radio stations, etc., external organizations (governmental or otherwise), who work with and/or within these communities; e.g. Ministry of Agriculture, research institutes, universities (research groups and experts), and local and/or external companies; e.g. traders, bulk buyers, transporters, Telcos, etc. A combination of the aforementioned form the shareholders of any ICT system in low-resource communities.

Together with the communities, these shareholders form an ecosystem that functions in its own way. As such, shareholder engagement is necessary to even begin to truly identify what the problems are in the specific context and what possible solutions could exist. In the case of one of the author's early projects, Mr. Meteo (Dittoh et al. 2020), the best the group of experts could propose was the fact that information would be useful to communities and the empirically backed fact that, mobile telephony and radio were the best technologies to consider. It then took a visit to the communities to ascertain the fact that climate information was a problem. In the case of Tibaŋsim, a platform was built for information dissemination to rural communities, but the manner of information was proposed by the community members; giving an indication of what is relevant to them, and the research institute; simply using the platform to improve the information they hitherto disseminated manually.

11.5.2 Indigenous Communication

Information and Communication Technologies, as the name clearly states, are meant to facilitate the transfer of information from person to person and place to place. Other African researchers have delved into the importance of indigenous knowledge (Millar and Dittoh 2005; Owusu-Ansah and Mji 2013; Briggs and Sharp 2004), but also indigenous communication is imperative for the dissemination of these indigenous knowledge as well as exogenous knowledge. In a BBC article[7] by Padraig Belton, a Technology of Business reporter, in which Mr. Meteo was fea-

tured, the writer asks an intriguing question; "*Will talking ever top typing?*". His article makes us realize that in most places typing never existed, how much more to take over talking. He goes further to mention how even in the most advanced countries, and in urban Africa, talking seems to be making its way back into the digital space; Amazon's Alexa, Apple's Siri, Microsoft's Cortana, Google's Assistant, young people sending voice-notes on chat apps, WhatsApp voice calls, Twitter Spaces, Zoom, Google Meets, etc. Why is this the case? I suggest here that it is because the indigenous communication method of all humans is voice, not text! This however becomes more relevant in the context of bringing exogenous knowledge to low-resourced communities. The developers of the mobile telephone did not engage in stakeholder involvement in Africa, as mentioned above, but the mobile phone accidentally met the requirement of utilizing the right indigenous communication method; the original communication method.

We must come to the realization that in communities, there is a local ecosystem, made up of *stakeholders* (see above), who have their *indigenous knowledge*, which is shared with their *indigenous communication* methods. Any ICT innovation or solution therefore must be "plugged-in" to the most appropriate place within this ecosystem.

11.6 The Plug-In Principle

The Plug-in Principle is a concept that is derived from lessons learnt from experiences in the area of Agriculture from Community-Oriented Education at the University for Development Studies (UDS) Tamale, Ghana. The principle is based on the following points:

1. Scientific knowledge cannot replace existing knowledge or situations, it only "betters" it. Therefore scientists, researchers and interventionists are at best "bettering" agents, not change agents.
2. The plug-in (intervention) should be "narrower" (have a minimal role to play) as com-

---

[7]https://www.bbc.com/news/business-43409952

pared to existing knowledge (in any society or community).

3. To successfully plug-in, there is the need to thoroughly understand the existing situation. Thus, researchers and interventionists need to spend time in the communities and with community members.
4. Community members will accept a message if they accept the messenger. The people bringing the intervention are as important as the intervention.
5. The amalgam of Indigenous Knowledge (IK) and Scientific Knowledge (SK) is very much dependent on the degree to which interventionists understand and appreciate the existing situation.
6. The understanding and appreciation of the existing situations help to modify intervention strategies to suit particular situations.

Building on the discussions of successes and failures and the reasons behind them; knowledge gleaned from years of ICT4D research, in conjunction with the plug-in principle (as outlined above), I present the Plug-in Principle for ICT4D;

Point 1. Modern methods of communication (namely ICTs) should not replace indigenous communication methods but should only better them. *As seen in the case of the Mobile Phone, the technology has made such an impact because it did not change the situation, it only bettered it.*

Point 2. The ICT intervention should be narrower as compared to the existing system. *Any intervention that is not narrower stands the risk of being too disruptive.*

Point 3. To successfully plug-in a new ICT system into the existing system, ICT4D researchers and developers must thoroughly understand the existing situation.

Point 4. ICT4D researchers, and perhaps even programmers need to be known, and accepted by stakeholders and community members as far possible. *African*

communities   are   largely people-centered.

Point 5. Our interventions must produce an amalgamation of a system that is mostly indigenous but integrated with our developed ICTs; this requires that we understand and appreciate the existing system.

Point 6. This understanding will aid to build systems that suit specific situations.

Point 7. One extra facet that ICT interventions may face is the replicability and scalability of our systems, since this is a major advantage of ICTs we would not want to lose. This requires an extra step in design and development where, as we develop systems in one use-case or for one community, we must consider that these systems, or at least versions of them, will be replicated and/or reused in other places or within other systems. This not only requires open-source, open-access development, but requires developers and researchers to actively brainstorm and consider the reuse and scaling of developed systems.

11.7 How to Plug In

11.7.1 What Is Plugged into What?

First things first; what exactly are we plugging? And into what? This question is vital and should be asked for every developmental intervention. In the case of ICT4D, it may be a bit easier than most fields to spell out exactly what we are plugging in; *Technologies* that will aid *Information access* (from the internet, any other data source, stakeholders, etc) and *Communication* (amongst community members and/or between community members and externals (other individuals, institutions, organizations, stakeholders, etc)). Perhaps the more interesting (and difficult) part of the question for us as ICT Experts is; plugging into what? We have our grand idea, our well-crafted IT solution, our carefully designed project; where exactly do we place it technically, and what role will it play socially?

Figure 11.1 shows us the various aspects of information and communication in communities; information flow within the communities, between members of the community and others outside; family, friends, acquaintances, traders, farmers, etc. that live in other places, information from, and to, media providers (TV, and especially radio), information to and from institutions and organization, both governmental and non-governmental, other digital information systems that may exists, and information to and from the internet; which of course is mostly non-existent.

More importantly, information flow is a composition of various constructs including, but not limited to; *the method of communication* (e.g. voice); *the technologies* being used, be it natural (hand, mouth) or man-made (phone, radio); *the people* (and communities, organizations) involved; *the language* (together with its uniqueness and nuances), and *the culture*; both of which can have a large influence on how the communication would take place; *the indigenous knowl-edge*, which is being shared alongside scientific knowledge in some of these flows of communication; and many other components we might not always account for.

By and large, the technology construct is where we wish to plug into within ICT4D, and as much as possible ensure that we do not influence these other aspects mentioned. Even then, it is important to focus on technology-in-use over technology-as-invention (Heeks 2009), which further minimizes the level of technology we are plugging in. As evidenced from my own projects, it does sometimes become necessary to introduce new methods and/or technologies but this again is minimized by gaining a good understanding of the rest of the existing aspects; these aspects form the context. Inherently, providing access to information results in access to exogenous knowledge, but it is imperative to not replace indigenous knowledge, but promote and augment. Fig. 1 therefore shows us that the intent of the principle is that which would have an intervention that is narrower than the existing system (Plug-in Principle Point 2), avoid replacing indigenous

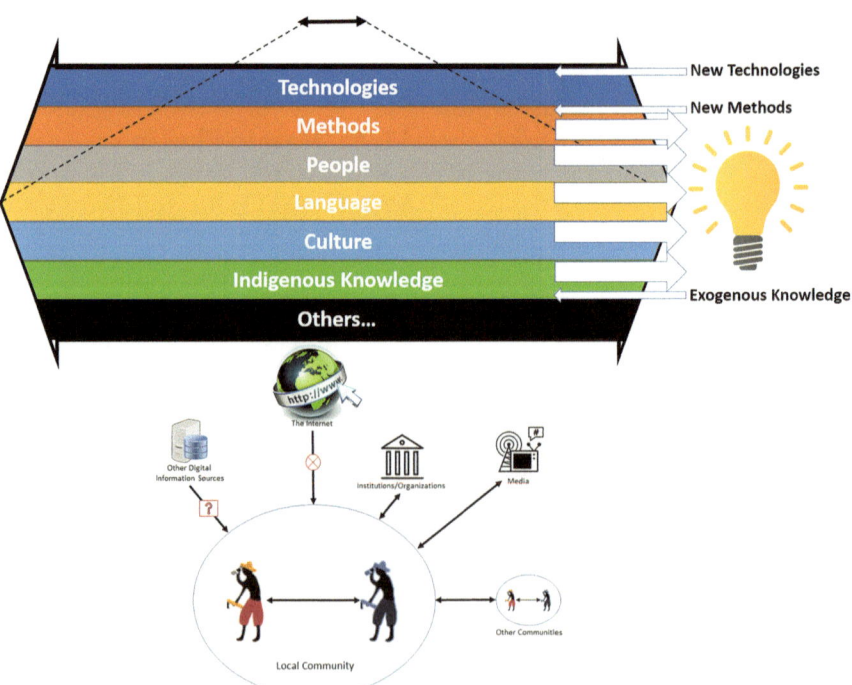

Fig. 11.1 The Plug-In Principle for ICT4D

methods (Plug-in Principle Point 1) and promote a good understanding of the rest of the system (methods, people, language(s), culture, IK, etc.; i.e. *context*) (Plug-in Principle Point 3).

11.7.2 Understanding the Context

Perhaps the most tedious step in the process is gaining a deep understanding of the context. Contextual issues are immensely important to solving any developmental issue and more so using ICTs. The Plug-in Principle therefore begins at this vital point; the need to have an in-depth understanding of the local context. Note that this cannot be done simply by remote observation alone but by active on-the-ground investigation of the socio-cultural context. Computer Scientists of course are not traditionally social scientists, but the argument here is that we need to become exactly that to a certain extent if we wish to understand the context in which low-resource communities we wish to provide services for function. The first thought you may have on reading this may be that a social scientist should handle this on your behalf, but unfortunately, no one can do this for us; the viewpoint we have as software and systems developers is very important to the process and yes, I have said that our intervention should be minimal as compared to the existing system, but that minimal "plug-in" is very vital and only we, having a good grasp of what ICTs entail can integrate that context with the local context, when we finally have a good understanding of it.

It is important to mention here, that in order to gain an understanding of local contextual issues, our own biases, however difficult, must be set aside, else they will influence our views on issues. This even becomes more vital because, without an intentional push to ignore our own bias, there is the tendency for local communities to "prefer" whatever technological artifact or scientific knowledge they believe to be developed or suggested by the "experts" as opposed to even identical alternatives. Dell et al. (2012) finds that when the interviewer is a foreign researcher requiring a translator, the bias towards the interviewer's

opinion or solution doubles. This also brings to mind the need for an increase in indigenous researchers, who would already have a bit more contextual knowledge and less external bias, in this field because a researcher's identity may enable access to a community (or prevent it) (Jimenez et al. 2021). A good (negative) example is my own bias in the early stages of becoming an ICT4D researcher where I assumed the introduction of an ICT system that cut out intermediary traders from the agricultural value chain was a good idea.

11.7.3 Understanding the Problem

As ICT developers and/or researchers, we cannot refute the fact that you may have an idea of the problems in the community or a target population, and you may even have the possible solutions. For example, we can confidently say that low levels of literacy are a problem in communities in the Northern Region of Ghana and we could even go further to say that building a school in one of these communities is a possible solution to this problem. However, as we have just seen, local context is of utmost importance, and is possibly different in different communities. The differences may be subtle or very diverse, but either way could mean very different approaches are needed.

I wish to posit that there are (1) perceived problems that do not really exist, (2) perceived problems that exist but are not considered as problems in the context of community members, (3) perceived problems that exist but are not considered relevant to tackle by community members and stakeholders, and (4) perceived problems that exist, are considered as problems and also relevant, by community members and stakeholders.

Due to the uncertainties of the classification of perceived problems as suggested, it is best from the onset to rely on the community and stakeholders to brainstorm on problems and even think of possible solutions (See Plug-in Principle Point 6). This is often neglected by researchers and developers mainly because, as we have seen from the various projects, it takes a lot of effort; workshops with stakeholders, discussions with experts, community visits, etc. When done well, the problems, as well as possible solutions (even if perceived beforehand), would be issues that are of relevance to the stakeholders, especially the community members.

11.7.4 Understanding the Solution

What is the solution to the problem we have identified? Who decides the solution?

This is a tricky one; there is the tendency to assume that illiterates and literates without ICT knowledge (digital literacy), knowledge of Computer Science concepts, or programming skills will be unable to offer ideas for ICT-based solutions. This is erroneous!! Granted, we as ICT developers must take lead on coming up with these solutions, but just as our contextual understanding results in an understanding of the problems, community and stakeholder engagement will result in the best possible solution.

One reason why this is helpful is the fact that we do not intend to build a solution that replaces their existing systems (See Plug-In Principle Point 1), but a system that integrates with what they know and do (See Plug-In Principle Point 5) and is as minimal as possible compared to the existing system (See Plug-In Principle Point 2). Engaging with local stakeholders and community members to come up with a solution naturally makes the above possible since they are involved from the very start.

Of course, the solutions proposed by stakeholders and community members, although absolutely relevant, must also be technically feasible and it helps to (during workshops and focus group discussions) explain (in lay terms) what is technically possible and what is not. It should not be expected that perfect solutions will result immediately; they may even be false-starts and the total scraping of proposed solutions, but the use of a good methodology, which is iterative in nature and based on feedback from evaluations from stakeholders and end-users will provide the ideal solution. (See Bon et al. 2016).

11.7.5 Building the Solution

Regardless of the methodology used, there should be constant community and stakeholder involvement; remember, we are "plugging in" our innovation into the community/stakeholder ecosystem to solve a problem they have identified together with us. The lifecycle of the solution development must therefore include end-user and stakeholder engagement, with prototyping where they can utilize the modules of your solution to see if they fit in with what they require and are inclined to use (See Plugin Principle Point 1 and Point 5).

One important thing to remember is the use of 'technology-in-use' (Heeks 2009); building innovations around technology that the communities already use, or if you must introduce new technologies, focus on simple, cost-effective technologies that can be used in tandem ("plugged in") into what they already use. An example is the case of Tibaŋsim where we introduce new hardware (Raspberry Pi) but we ensure it is low-cost technology, simple (communities only have to plug-in the device to power plug or power bank), and "plugged-in" to already 'technology-in-use' (Radio). This was the case with mobile phones as well where it was affordable technology, simple to use and "plugged-in" by way of its use of indigenous communication methods.

Also, just like there is the tendency to want to replace Indigenous Knowledge with Scientific Knowledge, which we must work against, but rather work towards integrating them (See Plug-in Principle Point 1), there is need to ensure that ICTs are as minimal as possible (See Plug-in Principle Point 2) to avoid disruption.

11.7.6 Sustainability and Scalability

Finally, what happens to these projects and interventions when we as scientists are done and have packed up and gone home? What happens when the funds for the project (which most likely came from some agency in the West) has run out? This has been a recurring theme of a lot of projects, including successful ICT implementations which did not fail for lack of following the right methodology, but simply run out of funding. It is therefore imperative that we as researchers and developers who are concerned, not only with increasing scientific knowledge and gaining empirical evidence of the various facets of the digital divide, but more importantly with using this knowledge to foster development, to consider what we can do, by way of design and implementation to make ICT projects financially sustainable (See Plug-In Principle Point 7).

If I were designing the latest AAA 3D Computer Game, I would most likely not be too concerned with the cost of the hardware required, based on the assumption that avid computer gamers have very high-end gaming machines and consoles. However, in designing solutions for low-resource environments, I would have to consider what hardware would run the implementation in terms of cost. This leads us to consider cheap but usable hardware, open-source software and platforms, minimal setup cost stemming from simple setup methods, and low running costs, so as to make sustainability of the project more feasible.

The issue of running cost then leads us to the concept of a business model; as ironic, counter-intuitive and controversial as it may sound, our ICT4D projects will remain pilots and inevitably degrade into disuse if they are not built around a business model that generates revenue. In his article, "*To Really Help the Global Poor, Create Technology They'll Pay For*" by Alex Deng, Chairman of Huawei's Corporate Sustainable Development Committee, in the Harvard Business Review, he makes the point that for many ICT4D projects to scale beyond a pilot, there needs to be some sustainability plan on how those projects will become self-sufficient in the long-term. Deng argues that we, as developers, need to produce solutions that the users will pay for and further argues that; "*supplying something that customers will pay for has another virtue: it provides feedback, which many digital inclusion projects lack. By charging even a small amount, ICT initiatives get a clear signal about how to adapt their offerings in response to changing conditions. Without this feedback, they operate in a vacuum of good intentions, insulated from the market and ultimately cut off from the very communities they are trying to serve. Nothing provides clearer evidence of a project's viability than a base of paying customers, and this proof of success makes scaling up vastly easier.*"

I fully agree with Deng in this; it is fundamental that, people value what they pay for and pay for what they value (Fay 1930; Szasz and Compassion 1998), as such it is imperative that our intervention must be of value (and if we have adhered to all we have learnt so far, then we know it is) and users must pay for this value. In addition to this, at least in the case of Tibaŋsim, but applicable to many other ICT4D interventions, is the addition of other revenue sources; stakeholder funding for services of value to them; inclusion of advertisement of relevance to communities by companies and organizations (sellers of seeds, agric-input, etc) and other forms of announcements (meetings, invitations, sales of produce).

11.8 Plugged In

Over the years, I have seen this principle, unintentional sometimes and very intentional in some interventions, in action in various fields, especially in agriculture and food security; having been involved in a lot of developmental work in rural areas mainly in the north of Ghana, I have seen many, many interventions over the years and unfortunately I have seen the many that failed. My experience has been that, all the failed interventions (ICT or otherwise) failed to adhere to the aforementioned principle. Those that succeeded for the most part followed these guidelines, mostly not as a scientific principle per say, but based on informed knowledge and experience, or simply by coincidence or accident. We can however not rely on accidents to close the digital divide; it must be intentional!! It is imperative that the Global South capitalizes on ICTs to effect a huge skip in development; we cannot afford to develop gradually, and the advent of technologies like the Mobile Phone, the Internet,

IOT, AI, VR, etc. has shown us that it is possible for technology to cause giant leaps in development.

We cannot afford failed projects and interventions at this stage; there's simply no time for that, and adherence to this principle in ICT systems for low-resourced environments will ensure their success, connect the unconnected and foster development through information and communication.

References

Baart A, Bon A, de Boer V, Dittoh F, Wendelien Tuijp W, Akkermans H (2019) Affordable voice services to bridge the digital divide: pre-senting the Kasadaka platform. In: Lecture notes in business information processing. Springer International Publishing, pp 195–220

Benjamin F (1999) Community development and democratisation through information technology. In: Heeks R (ed) Reinventing government in the information age. Routledge, p 17

Bon A, Akkermans H, Gordijn J (2016) Developing ICT services in a low-resource development context. Complex Syst Inform Model Quart 9:84–109

Briggs J, Sharp J (2004) Indigenous knowledges and development: a postcolonial caution. Third World Q 25(4):661–676. https://doi.org/10.1080/01436590410001678915

Brooke J (1996) SUS: a 'Quick and dirty' usability scale. In: Usability evaluation in industry. CRC Press, pp 207–212. https://doi.org/10.1201/9781498710411-35

Dell N, Vaidyanathan V, Medhi I, Cutrell E, Thies W (2012) "Yours is better!": participant response bias in HCI. In: Proceedings of the SIGCHI conference on human factors in computing systems. CHI '12. Association for Computing Machinery, New York, NY, USA, pp 1321–1330. isbn: 9781450310154

Dittoh F, van Aart C, de Boer V (2013) Voice-based marketing for agricultural products: a case study in rural Northern Ghana. In: Proceedings of the sixth international conference on information and communications technologies and development: notes - volume 2. ICTD '13. Association for Computing Machinery, Cape Town, South Africa, pp 21–24. isbn: 9781450319072

Dittoh F, de Boer V, Bon A, Tuyp W, André Baart A (2020) Mr. Meteo: providing climate informa-

tion for the unconnected. In: 12th ACM conference on web science companion. ACM. https://doi.org/10.1145/3394332.3402824

Dittoh F, Akkermans H, de Boer V, Bon A, Tuyp W, André Baart A (2021) TibaNsim: information access for low-resource environments. In: Proceedings of sixth international congress on information and communication technology. Springer Singapore, pp 675–683. https://doi.org/10.1007/978-981-16-2377-6_62

Etzo S, Collender G (2010) The mobile phone 'revolution' in Africa: rhetoric or reality? Afr Aff 109(437):659–668

Fay CR (1930) Adam Smith and the dynamic state. Econ J 40(157):25–34. issn: 00130133, 14680297. http://www.jstor.org/stable/2223637

Heeks R (2009) The ICT4D 2.0 Manifesto: where next for ICTs and international development? SSRN Electron J. https://doi.org/10.2139/ssrn.3477369

Jimenez A, Abbott P, Dasuki S (2021) In-betweenness in ICT4D research: critically examining the role of the researcher. Eur J Inf Syst 31(1):25–39. https://doi.org/10.1080/0960085X.2021.1978340

Latham R, Sassen S (eds) (2005) Digital formations: IT and new architectures in the global realm. Princeton University Press. ISBN 0-691-11986-4, ISBN 0-691-11987-2

Millar D, Dittoh S (2005) Interfacing knowledge systems: local knowledge and science in Africa. Ghana J Dev Stud 1:2

Owusu-Ansah FE, Mji G (2013) African indigenous knowledge and research. Afr J Disabil 2(1). https://doi.org/10.4102/ajod.v2i1.30

Rogers E (2003) Diffusion of innovations, 5th edn. Free Press, New York

Schuelke-Leech B-A (2018) A model for understanding the orders of magnitude of disruptive technologies. In: Technological forecasting and social change, pp 261–274

Sharma V, Rao KV (2000) Bridging the digital divide: information Kiosks in rural India. In: Rural India: achieving millennium development goals and grassroots development. Tettey, p 383

Szasz T, Compassion C (1998) Psychiatric control of society's unwanted. Syracuse University Press. isbn: 9780815605102. https://books.google.nl/books?id=2pduB22E43oC

Tettey WJ (2000) Computerization, institutional maturation, and qualitative change: analysis of a Ghanaian public corporation. Inf. Technol. Dev 9(2):59–76. https://doi.org/10.1080/02681102.2000.9525322

Tracey-White JD (2005) Rural-urban marketing linkages. An infrastructure identification and survey guide. Food and Agriculture Organization of the United Nations, p 96. isbn: 9251053871

Make the *BOT* Speak Your Language—Plugging-in Artificial Intelligence into Women Entrepreneurship in the Sahel

Marie-Lou David and Anna Bon

Abstract

This Chapter describes a participatory action research project that aims to integrate digital technologies and local know-how in rural Africa. The results of the project are threefold: (i) the context with all specific conditions is explored for the potential design of digital technologies (ii) a digital solution (a so-called *bot*) is effectively designed in close collaboration with local stakeholders, while plugging into their interest, knowledge and practices (iii) as a meta-result, an approach, how this type of action research could be done, is developed iteratively, and validated. The project can be considered an instance of the plug-in principle, as it blends two different knowledge domains: local farming and trade in rural regions with digital technology design.

Keywords

Non-timber forest products (NTFPs) · Women's cooperatives · West Africa · Digital technologies · ICT4D · Plug-in

M.-L. David · A. Bon (✉)
Vrije Universiteit Amsterdam, Amsterdam, Noord-Holland, The Netherlands
e-mail: a.bon@vu.nl

12.1 Introduction

In rural West Africa, small-scale agriculture is one of the principal economic activities. However, climatic hazards due to climate change make harvesting increasingly uncertain. This exacerbates the vulnerability of many rural communities to food shortages and economic hardship (African Smallholder Farmers Group 2010).

In West African countries, notably in Burkina Faso and Mali, for several years, innovative farmer groups have been exploring alternative ways to escape from food and nutrition insecurity (Reij and Waters-Bayer 2014). Exploitation of non-timber forest products (NTFPs) is one of the coping strategies of in particular women's cooperatives to increase resilience to food insecurity within the communities. The production of these products from trees has become the main activity of many small local cooperatives.

NTFP represents environmentally friendly and sustainable farming. Recent studies have shown high nutritious values of non-timber forest products in local food consumption (Ahenkan and Boon 2011; Maua et al. 2019; Chou 2018).

Still, despite opportunities, the women's cooperatives are facing challenges related to entrepreneurship, limited communication channels, difficult access to markets outside their communities and to the larger value chains of food supply in the country. Generally, the women

© The Author(s) 2025
S. Dittoh et al. (eds.), *Integrating Indigenous and Scientific Knowledge for Sustainable Food Systems in Africa*, Sustainable Development Goals Series,
https://doi.org/10.1007/978-3-031-85512-2_12

cooperatives lack a solid country-wide network of partners. They do not receive recognition for their important entrepreneurial achievements in the local communities. This hampers their access to markets and customers.

The practical question tackled in this chapter by a local NGO together with an international, interdisciplinary research team, is how to explore ways to support the women cooperative's work and trade for enhanced exploitation of non-timber forest products.

Interested in exploring ICTs potential for low resource environments, the research team tried to design specific digital tools to improve communication and trade for the women and their communities. This chapter describes how this has been done, what we learn from it and how this can be generalised for digital design in low resource environments in general.

The chapter is structured as follows: Section 12.2 describes the start of the case study on women entrepreneurship in the Sahel, focused on the production and trade of non-timber forest products. Section 12.3 introduces a methodology that has been previously used to develop digital technologies in rural regions of Africa in a collaborative, iterative, adaptive way. This method was coined ICT4D 3.0 and seems appropriate as a starting approach for this research as it is appropriate for open-ended studies with many initial uncertainties. Section 12.4 describes the context of the case study, the stakeholders, and their needs. Section 12.5 describes the use cases that came up during the discussions with the stakeholders in the given context. Section 12.6 describes the technical design of an application prototype for one of the use cases that were formulated. The design decisions and the deployment are described according to well-established software engineering practices (Wohlin et al. 2012). Section 12.7 gives an evaluation and a discussion and gives ideas for future research. Section 12.8 summarises the three main results as briefly mentioned in the abstract.

12.2 Designing Digital Technologies for Communities in West Africa: How to Start?

The first question that comes up, in this practically oriented action research project, is how to design and deploy digital solutions such that they will be genuinely useful for the women groups and their entrepreneurship in rural Burkina Faso. It is clear that the research, before any clear goals are set, and before design plans are made, must start with a thorough context analysis, as various research team members have no extensive experience with farming and /or trading in rural West Africa, before starting this research project. So, essential questions must be clarified first.

To start with, we ask w*ho are the stakeholders in the exploitation of non-timber forest products? How do the cooperatives work, how are they organised, what information do they produce or use? What are the tasks carried out by a cooperative? Who are their customers? What are the workflows and procedures? Are they formalised or implicit? Which communication channels are commonly used?* These are just a few questions that must be discussed in person with the stakeholders. Focus group discussions, interviews and conversations are a very good way to clarify these questions (De Boer et al. 2012).

Secondly, the communication channels are explored. This is relevant information for the design of information and communication technologies. *What is the state of the art of digital infrastructure in rural regions? How is the use of smartphones facilitating the access to information in low-resource, rural context? Moreover, how do people here commonly communicate? Which language do they speak? Are they familiar with text-based communication? Do they read and write? And moreover, what can users afford in financial terms? Would they be prepared to spend money to use a system that enhances their work and trade?*

Thirdly, we ask, *what would make an application usable for the local users who are unfamiliar with ICTs? Are there any current technologies that can provide a cheap and effective solution for automated communication? And if not, what is the best approach to design and build a digital system in the given context(s)?*

After having briefly explored the context and the users needs, two questions come up, positioned at two different levels. At the practical level the question is: what is the best (digital/technological) solution for a problem of the women cooperatives? At the methodological level, *how can such a result be achieved in a low resource environment and what does this case study teach us about digital design in a complex unfamiliar context, in general?*

12.3   Choosing a Methodology and General Approach

After the first brief exploration, we embarked on a design science action research project. To do this systematically, we have to select and describe a framework or methodology. For pragmatic reasons, we selected a practical approach, developed for collaborative ICT development in West Africa, named ICT4D 3.0. This method combines digital development, design science and software engineering with social and ethical considerations. Overall, it has a collaborative, iterative and adaptive approach (Chou 2018) This ICT4D 3.0 framework has been field-validated and consists of five phases, not necessarily sequential, but rather iterative, being: (i) context analysis including a stakeholder analysis (ii) needs assessment to find out what the stakeholders and envisaged users really want to have improved or solved (iii) requirements analysis, to iteratively design and build an appropriate digital solution/system (iv) sustainability analysis to make sure that the envisaged solution can be retained after the research (pilot) project phase is finished, i.e. there is a business case for it (v) engineering and iteratively improving and deploying the solution in the local context.

The (meta-level research) question we pose here is: is ICT4D 3.0 an applicable method and approach, does it need to be updated? Is this method applicable, given the new technological advancements and their social implications such as the rapid emergence of Artificial intelligence (AI), the recent improvements in the local digital infrastructure and mobile telephony networks, and the increasing use of smartphones and social media in Africa. In the next section we tackle the case of women cooperatives in West Africa who produce, transform and sell non-timber forest products. We do this by investigating the context, stakeholders and needs, and trying to explore the problem space to design a possible solution to one of the pressing problems of the stakeholders. To do so, we need to structure all the unstructured, messy and chaotically obtained information, by using a flexible, but systematic approach.

The action research project started in 2021 and continued until late 2024 in Burkina Faso, and—to a minor extent—in neighbouring country Mali. The project has been structured according to the various stages of the ICT4D 3.0 framework. These are described in the following sections.

12.4   Exploring the Context, the Stakeholders, Their Needs, Their Operational Goals

To truly grasp the context of rural trade in West Africa, the authors met (online) with local stakeholders to exchange ideas around their challenges. The authors interviewed different people: an ICT4D expert from Mali, radio station operators in Burkina Faso and Mali who closely collaborate with women cooperatives, as these resource persons have first-hand experience in the context. A workshop was held in the presence of women leaders of the women cooperatives. It was a precious resource to understand their goals and how they currently operate.

An online, smartphone-based group-chat was established with the cooperatives' members in which the researchers had access to understand their daily conversations about their work. We did a stakeholder analysis to find out who are the players in the value chain and a brief analysis of the value product. In other words, a question that also needs to be clarified is: what are non-timber forest products NTFP, and how does the local trade work, what do the women cooperatives do? As the initial idea was to make an application to improve sales for the cooperatives, we developers needed to understand more about the product, the producers and their specific needs.

12.4.1 Stakeholders: Women Cooperatives in the Sahel, Radio, NGOs, Customers

In accordance with the Organization for the Harmonization of Business Law in Africa (OHADA, https://ohada.org), women selling non-timber forest products NTFPs organise themselves in cooperatives. A NTFP cooperative is usually a women's organisation in which participants from the same or neighbouring villages join efforts to produce and sell a certain product. The organisational structure facilitates administrative implementation and enables the participants to sell in larger quantities to clients than when they would do it alone. So, when approached by a client who wants a large quantity of a certain product, a cooperative can more adequately meet the demand. Thus, they jointly make more profit and spread the work among themselves.

NTFP cooperatives focus mainly on just one or two types of NTFPs. The cooperatives have varying sizes, which typically can be between 20 and 100 members. In the case of the Sumbala project, the typical size is of around 30 women. When a woman moves to a new village, she will usually join the local cooperative. At the head sits a board of directors, elected democratically every two or three years. The board comprises a president, sometimes a director of communication and organisation and other members depending on the hierarchy decided by the cooperative. One person is appointed as contact for potential clients. She is responsible for relaying the demand to the cooperative and communicating with all members to try to meet the demand.

Every member has a mobile phone, but it is not necessarily a smartphone. The women possessing smartphones and having access to the Internet are often used to using WhatsApp as a method of communication. Everyone is also accustomed to using mobile money services which essentially acts as an online bank for many people in Burkina Faso. The cooperative women use mobile money to avoid carrying cash on themselves which can be unsafe. Moreover, they are familiar with the system, know how to navigate it and they trust it because their relatives from cities often use this method to send money. Even though many women have low literacy skills, they still know how to read numbers and type them into a phone (Karakara and Osabuohien 2022).

Organisation of the Cooperatives

The organisation or board of a cooperative has as its goal to effectively manage their members, stocks and finances to make a profit. Indeed, one of the advantages of grouping themselves was to be able to aggregate production and thus attract more clients for sales.

However, managing stocks and finances can prove to be difficult without an efficient information system in place. The operational goal of a cooperative organisation is to trade more NTFPs, which can be helped by attracting more clients, supporting members, or through more effective management of stocks or finances. They also have a need to be more connected to the NTFP market in general, especially other cooperatives dealing in similar products, in order to have a better understanding on setting prices.

Cooperative Members

A member of a cooperative aims at selling their products to increase their income and therefore independence. This can allow them to support their family. They are also looking for support from the cooperative, in learning new skills, building a community and a reliable network to sell their goods.

Apart from the women's cooperatives, there are also other stakeholders in the local context, who may concretely benefit from an envisaged digital system. The goal of conducting needs assessment is to capture what ICT solution could potentially help its users. In the case of ICT4D, people are from remote areas where they are not aware of the range of capabilities from ICTs. This caters for future users' hardware and digital skills.

The cooperative infrastructure is central to this research and constitutes the main stakeholder for this project. The authors made the distinction between cooperative organisations which represent the board of a cooperative and their members. This is to reflect the fact that they have different goals for the system.

Radio Stations

Radios play an essential role in rural West African communities. They are the main method for information dissemination, as having a radio does not need much electricity and signal reception (Bon 2020). Radio stations are heavily involved with the SUMBALA project. As privately operated organisations, they aim at making a profit through airing advertisements. They are also invested in raising awareness around environmental issues. Indeed, they have multiple programs centred around protected plants, organic farming and tree maintaining. NTFPs can be very beneficial for preserving nature, therefore stations have an interest in promoting the use of NTFPs as their cultivation helps conserve and develop forests.

NGOs

Non-Governmental organisations (NGOs) are also part of this project, as they often act as facilitators for community development actions in their region. This is the case of the NGO Reseau MARP 1. They are implemented in Burkina Faso and they focus on helping participatory

approaches for projects around (i) food security, (ii) natural resource management and climate change, (iii) decentralisation, (iv) training and networking of innovative farmers, (v) promotion of microfinance, (vi) disaster risk management. They are involved in helping Burkinabe women trade more NTFPs because it is heavily linked to (i), (ii) and (v).

Customers

The people who buy NTFP can be business people and resellers, but also consumers of NTFP. Sometimes big orders do arrive, e.g. 500 litres of shea butter. People from urban areas seem to appreciate NTFPs and would be interested in buying them because they are of higher nutritional quality and cultural value than mass produced products. In the city, people generally use social media to find interesting products and can get them delivered directly to their house by motorcycle. In the case of NTFPs from rural areas, this process is more complex to carry out. Indeed, the women do not all have access to a smartphone and internet, therefore, they cannot advertise their products on social media or they do not know how. Moreover, they come from remote villages, so the delivery is not as fast and easy as from inside the city. According to Amadou Tangara, who is an expert in NTFP and its trade, the bus network is extensive, even in the countryside, and generally, a bus will stop close to the villages. One can put packages in the bus, give the contact of the person picking it up and at the destination stop, the driver will check that the right person has picked up the correct packages. This is a reliable method for delivering packages from one destination to another without having someone travelling with the products, besides the bus driver.

12.4.2 Exploring the Value of NTFP

In Burkina Faso many products count as NTFPs. This includes honey, shea nuts and its derivatives (oil, soap, creams), moringa, sumbala, African locust beans that can be used for food or medicinal purposes, neem oil and many other products. The women producers of NTFP gather raw materials in the woods and sell them after transformation, for example from shea nuts they themselves produce shea butter, a consumable product for cooking. The harvest of NTFPs varies with the seasons and weather conditions, therefore they diversify the specific products they offer, over the year.

There are considerable income disparities between men and women in Burkina Faso. This issue is exacerbated in the rural context. Indeed, rural women have less opportunities to access income. They rely on the production of NTFPs and small farming activities. For them, NTFPs could be a key source of income. Amadou Tangara cited a previous project named RadioMarche (Bon et al. 2016) which helped women to increase their NTFP sales in Mali. When interviewing the women after the project was implemented, some expressed that they gained more independence and autonomy because they did not have to ask their husband for money anymore. This shows the validity of NTFPs as a source of income for rural women.

NTFPs are found and traded in rural areas. This means that the women come from remote villages, not necessarily connected to a road. Finding clients can be challenging in such situations. NTFP traders often went to weekly markets in neighbouring villages to be able to sell their products. However, the clients were the ones fixing the prices, often at a disadvantage to the sellers. Unfortunately, the public is still unaware or uneducated about all the benefits that NTFP consumption could represent. This represents a loss of income for the women and can be explained by the lack of education around the quality and benefits of NTFPs. After the RadioMarche project which advertised NTFP qualities on the radio, it was observed that clients were willing to travel further and to pay the prices fixed by the women for NTFPs. In some cases, the price for the product has doubled because the women regained control over it. This underlines the importance of raising awareness around NTFPs and their advantages, which can be achieved with the help of ICTs.

12.4.3 Problems Mentioned by the NTFP Cooperative Members

The following problems and challenges have been identified during the interviews with members of the cooperatives:

- Marketing: The cooperatives have difficulties in reaching new clients and advertising the NTFPs they have available.
- Societal traditions: As explained by Jean Arnaud Sawadogo, in some regions of Burkina Faso, women traditionally hold a quiet role in society, where they do not often speak up and let men take public space. He observed that for some cooperatives in Passore, it is difficult to advertise themselves because they fear the judgement of others for appearing and speaking publicly, something they are not used to.
- Storing system: They have no storing system or storing management system in place so they cannot yet group their stocks in one place to facilitate meeting bigger demands of clients. This generates delays when gathering everyone's stocks and means that they can lose on business.
- Inventory management: In the same vein, the members of one cooperative can be spread out in different villages, so it is difficult to keep track of the quantity of product each member has.
- Setting prices: The cooperatives struggle with pricing their goods, especially when they are not aware of the market prices of NTFPs in other areas of Burkina Faso.
- Predicting production and demand: It is difficult to predict the production that the exploiters will have for one period because it depends on the weather and quality of the raw material. The women have a hard time responding correctly to the market, meaning producing enough but not too much, also because they face fluctuating demand. Furthermore, if a demand is too big for a cooperative, they have to decline it.
- Lack of treasury funds: To be able to make NTFPs, cooperative members might need ini-

tial investments. However, microcredit rates are too high for them. Their way of obtaining liquidity is by each investing 200 fCFA per week, the member who needs money can borrow it from the common fund and reimburses it with an interest between 1 and 5%. The money from the loan stays in the cooperative, so other members can also reuse it.

- Unequal methods of production: There is no archive or resources to reference the production process used when processing NTFPs. This creates a disparity in the production processes, especially when new people arrive and use their own methods. Therefore, the quality of products in one cooperative can differ, depending on which member made them.
- Middlemen: Middlemen sell NTFP from rural areas in cities. They add no value to the products and only reduce the margins that the women would make on the sale of the product. They also jeopardise the quality of the products, to sell more at a lower cost, for example by diluting honey or mixing multiple types of butter but only showing the higher quality label. Overall, this chapter explores the setting in which the research takes place. It enabled the author to isolate important challenges that the cooperatives face and those will be the basis for designing the system as well as the skills that the women present and the available communication structures that they have.
- Political insecurity: In 2022, two coups d'etats took place in Burkina Faso, generating a lot of political insecurity (Heeks 2017). Since the start, thousands of people have been displaced, having to leave their homes to escape the escalating violence by religious extremist groups and the military state (Dittoh et al. 2022). As a result, travelling has become unsafe, especially in the northern and eastern part of the country. More people need extra income to survive but the insecurity makes it more difficult to travel and sell products in some regions. Furthermore, the phone lines can sometimes be disconnected by armed groups. According to our interviewees, the situation seems to be getting safer in 2024, but is still considered unstable.

12.4.4 Opportunities and Challenges Mentioned by a Resource Person on NTFP

After a discussion with an ICT4D and rural development expert from West Africa, Amadou Tangara, more specific needs have been identified which can be answered through use of ICT.

In remote areas, where the internet connection is sparse and people do not all have smartphones, communication can be difficult. In this case, it can take place at different scales. The smallest scale is the struggle for members inside one cooperative to share their stocks, decide where and how to meet. One need for a cooperative is to have access to a way to reach all members easily at one time. It would make it easier for them to collaborate on sales and trade. ICTs in that case can be used to connect members through a voice application that surveys them on what their stocks or availability are for example. As they all have phones, this could be achieved.

On a bigger scale, cooperatives struggle to make themselves known to potential clients. Indeed, people are becoming more and more connected to social media, but the cooperatives are lagging behind. This causes them to lose out on potential clients. ICTs can be used in multiple ways to improve advertisement to people who are already present. This could be on social media for example. There is a need for more advertisement, and this can be achieved through communication technologies.

Lastly, an important part of knowledge sharing is to learn more about one's craft. ICTs can be leveraged to enable a nationwide sharing of best practices in terms of collecting and producing NTFPs. A forum could be helpful for everyone in the NTFP community to share their experience and learn from others. This approach has been tested previously in various countries for farming and it proved to be beneficial to farmers because it improved their skills and strengthened community feelings (Morten et al. 2022; Eizenga 2023).

In addition to providing access to information, ICTs provide solutions to store information for future use and to build a knowledge base for one's organisation. In the cooperatives' case, it could prove useful to document their work processes in order to harmonise them and easing newcomers into their practices. Another way to leverage a knowledge base for them would be to record the demand for certain products over time to try to predict and answer it the most accurately possible.

Keeping records is an essential part of examining one's operations to improve them. On a bigger scale, one can imagine a nationwide knowledge base for prices, where each cooperative can record the latest prices goods are sold at. This could help every cooperative to ensure that their prices are coherent with the rest of the offer on the market. Currently, cooperative boards have no real access to price history for their sales. They are struggling to determine the ideal price of their products. A price knowledge base that they could access at any time and update would alleviate this problem.

The above section covered the goals of the different stakeholders for this project. And, it highlighted the general technical needs. For the rest of this research, it is essential to keep in mind the need for improved communication, within a community or outside of it.

12.5 Use Case Analysis

Once the context and the needs of the future users are well understood, specific use cases can be selected, elaborated and presented to the relevant stakeholders for selection. In order to properly develop one of the use cases, non-functional requirements are decided on. From the context analysis and needs assessment, multiple use cases emerged. They are described in the following section, and the system's main functionalities will be derived from these use cases.

12.5.1 Use Case: Radio Advertisement

Current situation: Potential buyers of NTFPs are often unaware of the offer that the cooperatives

Table 12.1 The different actors of this system and their respective goal

Actor	Operational goal
Cooperative	Advertise their products to sell them
Radio station	Play advertisement in exchange for money
Potential clients	Listen to the radio and potentially buy NTFPs from local cooperative

have. The cooperatives currently do not advertise the products they make and it is difficult for the general public to find them if they want to buy specific products. Moreover, in some regions of Burkina Faso, women can be shy to appear on the radio and make their voice heard to their close community.

Key idea: Broadcasting on the radio reaches many people in Burkina Faso. Having a system that automatically generates messages on the radio to advertise the products that the local cooperative offers and who to contact to get them, would increase the business of said cooperative. It also alleviates the fear of certain women to be a public figure on the radio (Table 12.1).

12.5.2 Use Case Processes Repository

Current situation: Members of cooperatives have all their own way of processing NTFPs. This creates heterogeneity in the quality of the products they offer, especially if multiple members group to meet a bigger demand. When a new member joins the cooperative, there is no set process in place to teach them, thus, everyone employs different techniques.

Key idea for an ICT solution: The key idea is to create an application that acts as a repository for each cooperative to store and access their processes. This will help the uniformization of methods inside one cooperative. When a new member joins, the other members can teach her their ways of doing things using the database as a resource. This can also help sharing techniques between cooperatives so they can uplift each other and learn more (Table 12.2).

Table 12.2 Actors and their goals for the use case of processes database

Actor	Operational goal
Cooperative member	Record +access process of making an NTFP
Cooperative coach	Train new recruits with the help of recordings
Cooperative	Exchange with other cooperatives about processes of fabrication

Table 12.3 Shows the different actors of this system and their respective goal

Actor	Operational goal
Cooperative—member	Make their stock or availability known
Cooperative administration	Centralise the stock or availability knowledge of the cooperative member

12.5.3 Use Case: Stock and Meeting Survey

Current situation: Members of one cooperative are geographically split, sometimes living in different villages. Communication can be difficult between all the members. This is especially a problem when a client approaches the cooperative with a bigger demand, the point of contact has to call everyone individually to be aware of the stock available.

This delays the sale. The process of organising a meeting is similarly complex because there is no centralised method of communicating with every member.

Key idea: The key idea is to create an application enabling the administration of the cooperative to quickly survey the stocks available of each member or to consult them on their availability for a meeting (Table 12.3).

12.5.4 Use Case Web Shop for Urban People

Current situation: People living in urban areas have no means of knowing about the NTFP offer in rural areas and of ordering them. However, they would be interested in buying them because the products are perceived as high quality goods when they are handmade. This represents a loss of business for cooperatives.

Key idea: City dwellers have for the most part smartphones and are used to finding different products on social media. Therefore, the idea would be to create a web shop advertised on social media, which presents the NTFPs available at a certain cooperative.

Table 12.4 Actors and their goals for the use case of a web shop

Actor	Operational goal
Cooperative	Register their products to sell them Client Order NTFPs

The web shop would give the clients the opportunity to order goods directly from the internet, they can pay with mobile money that the women in cooperatives trust, and the women can ship the products on the bus (Table 12.4).

12.5.5 Use Case: Artemisia Advertisement

Current situation: Farmers are starting to grow artemisia for its health benefits (see e.g. De Ridder et al. 2008). However, they struggle to sell all of their products. This is probably due to the fact that many people are not aware of the benefits that artemisia has for curing diseases like malaria.

Key idea: Similar to the first use case, using radio to broadcast messages would be useful in creating awareness of artemisia benefits and of farmers selling them. This would increase the potential client base of the farmers and help them make more sales (Table 12.5).

Key Quality Attributes

A key part of designing a system involves identifying the qualities it needs to fulfil in order to be useful for its users. From the context analysis, emerged three main quality attributes. The attributes described in the following section are essential for the solution developed to be suc-

Table 12.5 Actors and their goals for the use case of a web shop

Actor	Operational goal
Artemisia farmer	Sell artemisia products
Client	Hear about benefits of artemisia products and potentially buy some
Farmer umbrella organisation	Defending interests of small-scale and family farmers

Table 12.6 Main quality attributes of the system

Name	ID
Financial viability	QA-FINANCE
Usability	QA-USABILITY
Language accessibility	QA-LANG

cessful. They will drive the selection of the use case and furthermore the design decisions. They are summarised in Table 12.6 and explained furthermore in the rest of the section.

Financial Viability
ID: QA-FINANCE

Description: The system is affordable for every stakeholder. This includes members of cooperatives who do not have the means to maintain an expensive system. Therefore, it should be very inexpensive for them, while creating added value, so that the system stays viable in the long term.

Rationale: This project benefits from funding from the Dutch government and universities, so we have the initial investment to create a solution. However, the funding will stop eventually but the project should be affordable enough for all parties that when the international actors pull out of it, it still can function. Indeed, this project aims at improving the marketing of NTFPs, and if done successfully could improve income of poorer rural Burkina. communities. A previous project, RadioMarché, was implemented in Mali and proved useful but it was too expensive to keep it running when the funding stopped.

Ultimately, the goal here is to generate long-term and sustainable impact in Burkina Faso, so financial viability is essential for success.

Usability
ID: QA-USABILITY

Description: The system is easy to install and use for every stakeholder. The main stakeholder of our system is the women in the cooperatives who sell NTFPs. It is important that the system is simple to use, straightforward and developed with people from low (digital) literacy background in mind. Thus a user-centred development (Mathieu N'dri and Kakinaka 2020) will be used in order to develop an application that best corresponds to the abilities of its users.

Rationale: The ICT solution developed in this project is destined mainly for women who have low literacy and digital literacy skills. They are averse to change and might struggle to adopt a new system if they do not understand it well. For the application to be successful, it is essential that the system is correctly adapted to its users and to limit its learning curve.

Language Accessibility
ID: QA-LANG

Description: The system supports multiple languages most relevant to its users. In our case, it must include French, Moor., Dyula, Bissa and Fula.

Rationale: More than 60 languages are spoken in Burkina Faso so not everyone will have one language in common. The language accessibility of the system is necessary for it to be adopted by as many people as possible.

12.6 Developing, Testing, Deploying

Once the context, needs and use case analyses are done, we can start designing the solution. It is important to correctly plan out and consider the expectations from the system. Moreover, this section maps out the functionalities of the solution, to also showcase what it will not be able to perform. The following chapter expands on the system. First, a use case is selected from the ones which emerged during the context analysis. Then this chapter goes in depth about the designing phase of the system, its development and finally deployment.

12.6.1 Use Case Selection

When looking at the five use cases we identified for this research, we wanted to select a use case which involved radio, as this is an important channel for communication. This was only valid for use case 1. Moreover, the radios have signed a convention to help cooperatives with their marketing strategies. As radio workers or people from local NGO are proficient technology users, they can support the women in the cooperatives in navigating the system.

In addition to implementing use case 1, we saw the opportunity to combine it with use case 3 in order to design a real pipeline between surveying stocks available and directly generating.

a communique about said stocks. This solution would relieve two needs of cooperatives, the one for a more organised way of collecting information and the one for easier communication. For a better understanding of how the system is expected to work in the typical use case, a scenario is pictured in Fig. 12.1.

Use case 5 is a special case because it came later in the thesis project. It was presented to the author when she had already started developing the prototype. It was decided that the current chosen solution could fit the needs of Artemisia farmers.

This section is mostly dedicated to the design of the solution for NTFP cooperatives because they will be primary recipients. However, everything will be translated in Bambara for Malian users and adapted to the Artemisia use case for further testing. The first step to use the system is for each cooperative member to input their available stocks of NTFPs. When the cooperative administration wants to be advertised on the radio, they can trigger the generation of a communiquer in relevant languages. The communique is then sent to their local radio station which can broadcast it on allocated time slots. When the radio listeners hear the communique, they are made aware of the cooperative and can contact

them if they are interested in buying some products, thus generating profit for the cooperative.

12.6.2 Functionalities of the System

After establishing the general scenario of the system, it is important to consider what functionalities it will be expected to perform. The functional requirements are listed in Table 12.7.

The categories are mapped to actors in Fig. 12.2, for a visual representation of who does what in the system. It is important to note that the radio station is connected to all use cases related to the communique and the administration board. That is because the radio workers or people from the NGO Reseau MARP are the ones who will use these systems day to day as the cooperative members and administration do not have the necessary skills to. This is further described in the following section. This representation enabled us to isolate the system into three main components who will manage respectively: the generation of the communique. (purple), the administration board (red) and the stock survey (yellow).

12.6.3 Design Decisions and Requirements After Deliberation

Now that we can split the system into three components, responsible for different tasks, we need to compare options on how to best carry out those considering the quality attributes defined during the context analysis. These decisions were helped with an ICT4D expert (Mr. Stephane Boyera) in designing similar systems.

- Administration board: This component of the system is for the use of the admin person to manage who can input stocks and what types of NTFPs they provide. However, the women making up the administration board might not be skilled enough digitally to navigate a system with so many possibilities. A way to circumvent that is to delegate the administration of the system to Reseau MARP. Indeed, they

Fig. 12.1 General
scenario of the use of
the system

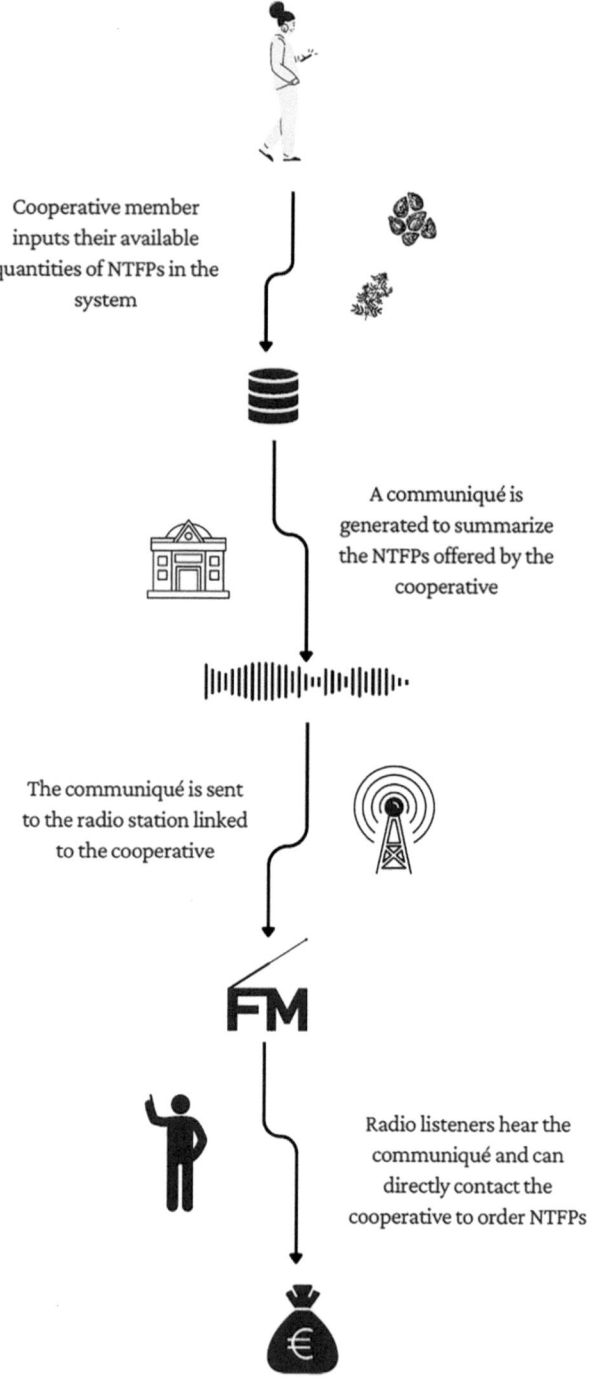

Cooperative member
inputs their available
quantities of NTFPs in the
system

A communiqué is
generated to summarize
the NTFPs offered by the
cooperative

The communiqué is sent
to the radio station linked
to the cooperative

Radio listeners hear the
communiqué and can
directly contact the
cooperative to order NTFPs

have more experience with ICTs and are liter-
ate. This simplifies the task of designing a
usable system.

- WhatsApp bot: This option aims at communicating with an administration member through

a bot on WhatsApp. This fulfils
QA-USABILITY because the people in
Burkina Faso are familiar with the app, they
will have the necessary hardware and we will
provide relevant languages for QA-LANG.

Table 12.7 Functional requirements for each use case

Category	Name	Identifier	Description
FR-STOCKS	Input stocks	FR-INPUT-STOCKS	A cooperative member is be able to list the quantities they currently have available of key NTFPs.
FR-REG	Register member	FR-REG-MEMBER	A cooperative administration is able to add a member to their structure so that the system can contact them.
	Register NTFP	FR-REG-NTFP	A cooperative administration is able to pick their 3 key NTFP that they specialize in.
FR-CONSULT	Consult stocks	FR-CONSULT-STOCKS	A cooperative administration is able to consult the available stocks of the cooperative.
	Consult list of NTFPs	FR-CONSULT-NTFP	A cooperative administration is able to access the list of their registered NTFPs.
	Consult list of members	FR-CONSULT-MEMBER	A cooperative administration is able to access the list of their registered members.
FR-DEREG	Deregister NTFP	FR-DEREG-PRODUCT	A cooperative administration is able to delete a NTFP from their list.
	Deregister member	FR-DEREG-MEMBER	A cooperative administration is able to delete a member from their list.
FR-COMM	Add contact information	FR-EDIT-CONTACT	A cooperative administration is able to add and modify the phone number on which they want to be contacted for sales.
	Generate communiqué	FR-GEN- COMM	A cooperative administration is able to automatically generate communiqué from the system info on stocks.
	Send communiqué	FR-SEND- COMM	A cooperative administration is able to send a communiqué directly to radio stations for broadcasting.

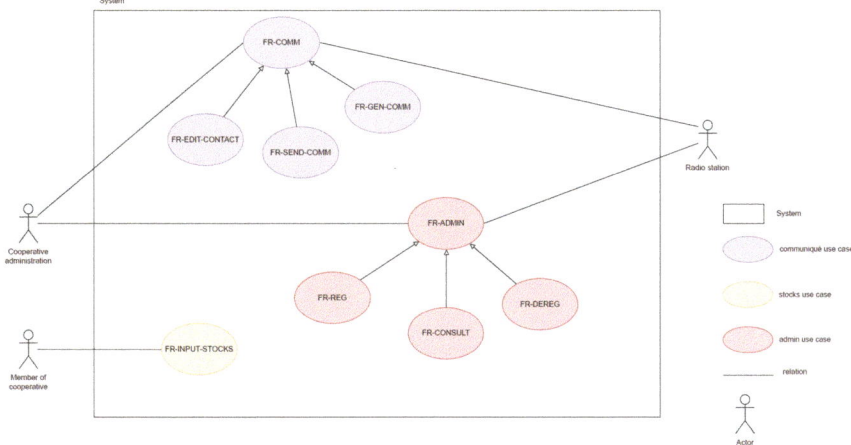

Fig. 12.2 Use case diagram: who does what, using the system, This enabled us to isolate the system into three main components: the generation of the communique, (purple), the administration board (pink) and the stock survey (yellow)

However, in order to develop a bot on WhatsApp, the business API is needed and it is not free. Moreover, it would be a complicated system to navigate with 17 cooperatives and at least 10 members perm cooperative; it does not fulfil QA-USABILITY. This option is not decided, because it fails QA-FINANCE and QA-USABILITY.

- Voice application: This option makes use of a voice application instead of a bot. It fulfils

QA-USABILITY and QA-FINANCE, however, other options can handle connection troubles and asynchronous messages as well as a large influx of information. Therefore, this option is not decided.

- Web application: This option makes use of a web application instead of a bot. It fulfils QA-USABILITY and QA-FINANCE because workers of reseau MARP know how to navigate a website. Moreover, this option is more adapted to access a lot of information. Indeed, the NGO might manage multiple cooperatives, each with up to thirty members who each have different stocks. This information would be better displayed on a web page. Therefore, this option is decided.
- Stock survey: This component is centred around having cooperative members fill in and update their NTFP stocks. Choosing the appropriate method of collecting data is essential for this component to work well. Let us present the three main options and select one based on quality attributes.
- SMS: This option involves querying the member to update their stock via SMS. However, as most cooperative members are illiterate, they are not used to SMS technology and will struggle to input their stocks. This option violates QA-USABILITY, therefore it is not decided.
- Voice application: This option makes use of a voice application to avoid using text. It has been validated before on systems such as RadioMarché (Bon et al. 2016) or Avaaj Otalo (Morten et al. 2022), which proved to be quite efficient, therefore it would fulfil QA-USABILITY on condition of being carefully developed. Moreover, this solution requires funds to get access to a dedicated phone and maintenance, so it does fulfil QA-FINANCE. Due to the political instability in the country and the general patchy phone coverage, this option relies too much on phone lines. Indeed, once the call disconnects, all data is lost, so it might not be a very reliable solution, so it is not decided.

- Bot on messaging application: Online messaging applications such as WhatsApp are gaining a lot of traction in the world and especially in West African countries where it is now considered cheaper to use than pay for a mobile subscription if you have access to the internet. Therefore, a bot developed on a messaging application would fulfil QA-USABILITY as it is a familiar application. Moreover, on applications like Telegram, developing a bot is completely free of charge, so this option fulfils QAFINANCE.

Finally, because the communication is done through messages, patchy internet does not prevent the bot from functioning. Telegram manages by itself messages sent without connection and stores them to send when the phone is back online. The conversation can take place over multiple days if needed without its flown and relevant information being lost in translation. Therefore, this option is decided.

12.6.4 Prototype

Due to the political insecurity in Burkina Faso in 2024, no workshops took place with the cooperatives in the country to test the working prototype during this research project. Therefore, the researcher/developer chose to focus on implementing only some parts of the proposed solution. From the design decisions above, a relatively new technology was chosen: Telegram[1] bots for interaction with the heads of cooperatives. This was settled on because (i) each cooperative has been given a smartphone during the Sumbala project implementation and (ii) they are familiar with messaging applications such as WhatsApp (Prasad et al. 2021). This research proposes an interactive bot on Telegram in order to test its usefulness in a rural area of a developing country.

---

[1] *Telegram* is a cloud-based mobile and desktop messaging app with a focus on security and speed. See https://web.telegram.org

The administration board is made for the Reseau MARP to be able to manage cooperatives.

Administration

In an effort to keep the system as usable as possible, it was kept extremely simple. From the context analysis, it was gathered that cooperatives typically focused on making a few key products. Therefore, to limit choices in the system, a cooperative can only keep track of up to three products from the following list, which was made after consulting the heads of cooperatives. They are deemed the most common types of NTFPs produced in Burkina Faso.

- Balanites juice
- Balanites leaves
- Baobab bread biscuit
- Baobab juice
- Baobab powder
- Herbal tea
- Honey
- Moringa powder
- Neem oil
- Shea butter
- Shea nut
- Soap
- Soumbala
- Zamenin

The products should be pre-entered by the admin, e.g. a staff member from the NGO, for each cooperative. Then each member can interact with the bot on Telegram to input their stocks of each product. Once they are done, a communique is automatically generated and sent to them so that they can share it either with a radio station to be broadcasted or to different group chats for advertisement.

The database and corresponding admin board have been implemented in Python using the Django framework[2] (Patel et al. 2010). It was chosen for its efficiency in getting a project up

and running, creating a managing interface for admin users and setting up a database.

12.6.5 The Bot Speaks to the User

A user accesses the *bot* through the Telegram application and can send any messages to get a reply. When messaging the bot for the time, it will ask for a phone number as they are not directly available through the Telegram API. If the number is in the system, the bot will give a list of the member's products and respective quantities. It will prompt the user for which product quantity to edit. As shown in Fig. 12.3, the bot generates a button with the corresponding products, and the user just has to click on the relevant one.

Next they are prompted to write down the quantity of the product they currently have. It is solely needed to write numbers, which is doable even for people with low literacy skills. Indeed, from the context analysis, it was gathered that they know numbers usually, especially to use with new money. Once the user has entered a correct amount, they can confirm or abandon the

Fig. 12.3 A screenshot from the smartphone with Telegram and the voice fragments generated by the bot

---

[2]*Django* is a high-level *Python* web *framework* that encourages rapid development and clean, pragmatic design. See https://djangoproject.com

interaction and restart from the product selection. If they confirm, the quantity of the relevant product is modified in the database. If the user is done with changing quantities, an audio message is sent which corresponds to a communiqué. This message contains the names of the cooperative, what products they offer if a member has some in stock and the number at which to contact them if someone wishes to make an order.

The state transition diagram of the bot is shown in the Appendix. Every time the user enters an incorrect input, or if an error occurs internally, the bot notifies them, as shown in the diagram.

12.6.6 The Use of Artificial Intelligence

Artificial intelligence (AI) is currently undergoing a huge developing phase worldwide. Large language models are yielding better and better results and companies are racing to surpass themselves. However, in order to be trained, models need enormous amounts of written or audio data in a language to become proficient in it. For very common languages such as English or French, the data is easy to find online. However for languages spoken in countries of rural West Africa, recorded resources of languages for Mooré for example are more difficult to come by. AI could be enormously successful in helping build systems accessible to communities with low literacy skills.

In 2023, a project to preserve world language with AI, named the Massively Multilingual Speech project (MMS) was launched by Meta.[3] This project aimed at extending AI models for speech processing to languages that are often forgotten from mainstream models. They successfully scaled models on more than a thousand languages, using recordings and transcriptions of religious texts. Notably, for Mooré, they created Text-To-Speech (TTS) and Automatic Speech Recognition (ASR) models. These could be very

useful for the use case, as its aim is to fully implement voice functionalities to facilitate navigating the system for users with low literacy.

In this research, the author proposes to introduce some of these currently available Text-to-Speech functionalities to the bot, but not exclusively. Future users are not used to artificial voices and they might be surprised by its use. This also aims at testing the accuracy of the model with different input. This research is a great opportunity to try a newly developed model and test its usefulness, while maybe finding some of its limitations. ASR functionalities are deemed too error-prone to implement in a system developed in a short time span without the possibility to really test it in real conditions.

For the use case at hand, it was decided that some outputs from the system will be recorded by a native Mooré speaker, to cultivate familiarity in the system, especially for the first voice messages heard by the user. It has been found very important to use familiar voices to boost usage of voice application (African Smallholder Farmers Group 2010; Eizenga 2023). A few key outputs were selected to be spoken through Text-to-Speech and this includes outputs with variable text. For example, the list of products and their respective quantities will be produced by the AI model. All fixed output, for example asking the user to click on the relevant button or error messages are pre-recorded audios.

The only exception made is for the communiqué, which is technically variable output; it depends on the cooperative and its current products available. However, because it is destined to other people, especially to be broadcasted on the radio, the artificial voice is not deemed authentic enough. Therefore, recordings of key sentences of the communiqué, names of cooperatives, NTFP names and numbers (for advertising phone numbers) have been made in Mooré, to assemble a custom communiqué.

The MMS models of Meta do not include numerals. Therefore, an important part of preprocessing text for the TTS model is to write out numbers. For common languages, Python libraries exist for that purpose, however it is not the

---

[3]Meta, https://meta.com is the big technology company that owns Facebook.

case for Mooré. A contribution of this thesis is a class method which takes as input an integer and returns the written-out number in Mooré.

Regarding the Artemisia use case, Meta's MMS project does not include the Bambara language. No other TTS model was found which performed reasonable accuracy. Therefore, the implementation in Bambara will not include AI. This can be seen as the control subject, compared to the Mooré implementation. It is essential to compare users' reactions towards a bot that uses AI and one which does not.

12.6.7 Deployment

As of now, the bot is deployed on PythonAnywhere 1. This platform provides a free tier for hosting services and since the database and bot do not use many resources, thus the free tier is sufficient for deployment. The only condition to keep the web application running on PythonAnywhere is to click a button to keep it activated at least every three months. This ensures that they do not have idle applications uselessly consuming their resources. The bot is accessible on Telegram by looking either for CommPFNL when wanting to access the NTFP bot for cooperatives or VenteArtemisia for the Artemisia version of the bot. For the users to be able to interact fully with the bot, they must be registered on the admin panel.

12.7   Evaluation and Discussion

The chapter presents the findings from elaborating the prototype. Unfortunately, due to the political insecurity in Burkina Faso and Mali, the team of project Sumbala could not travel to meet with its participants and help them set up the bot before the end of the research period. This would have been an opportunity to further explore possible design flaws of the system and iterate on development to improve them. However, this chapter reflects on the prototype created and its long-term sustainability. Moreover, it provides insights on the choices made to build it, the chal-lenges encountered and potential for improvement.

A prototype can only be validated through users in the field who can show proficiency in using the system and attest to its usefulness for their everyday lives. Because this was not possible at the time of writing this thesis, this section explores the sustainability of the bot and the challenges of its development.

12.7.1   Sustainability Analysis

An essential part of the ICT4D3.0 framework is the sustainability analysis. It aims at ensuring that the project can be successful long term. It is important to carry it out in the early stages of a project to verify that it makes sense to implement the project for all parties involved. One way to verify that is to ensure value co-creation (Chengalur-Smith et al. 2021) which assesses the added value of using the system for each stakeholder.

In the case of NTFP commercialisation, the stakeholders are: cooperative heads and members, Reseau MARP, radio stations and potential clients. The cooperative heads and members both stand to gain time and money by using the system. They gain time by having a streamlined process to enter their quantities and make advertisements. Furthermore, they could increase their sales with said advertisement. Reseau MARP's stake in the system is their goal as an NGO to help local communities take action to improve their business.

Radio stations benefit from the system by having advertisements already made, so they gain time and they get paid to broadcast them. At first, project Sumbala had a budget set aside to subsidise ads on the radio, but if the cooperatives notice an increase in sales due to advertisement, they will pay themselves for it eventually. Lastly, potential clients get to hear about NTFPs and in exchange for money to cooperatives, they obtain nutritional or medicinal products.

The system was developed for free during a master's thesis and the smartphones for each cooperative were purchased through the Sumbala project. Therefore, these costs are not a problem anymore. The bot, administration panel and database are all hosted for free on PythonAnywhere. It will stay hosted there on the condition that the owner of the account (the author of this thesis) clicks on a button every three months to ensure that the web application stays active. The authors plan on keeping it active as long as the system is used. Moreover, the code is open source on Github 1, so the project can be deployed again somewhere else, should Pythonanywhere disappear.

The costs incurred from the system itself would stem from maintaining or scaling it. Maintenance or implementing new features would require a developer. The payment could come from the cooperatives if they make enough money from sales of their products. A problem that could happen in the future, if too many users join the system, is scalability. Indeed, the database is a simple SQLite database, so if the number of users grows exponentially, it could become a bottleneck. The same thing goes if too many people use the system at the same time, as the server is also limited by Pythonanywhere free tier. This project so far focused on making a solid proof of concept and scalability was not a concern when developing it.

Developing a bot on Telegram was relatively easy because of the extensive documentation available online, which is also being kept up to date as it is maintained. Moreover, it is simple to deploy and be accessible to everyone with the internet due to platforms for low-resources projects. Telegram also manages messages sent and received when the device is offline, thus a patchy network connection is not a problem to use the system. This is a real improvement from voice application on a phone line, while keeping a similar flow of interactions on a platform that is familiar to the user.

Using the AI model trained by Meta was also relatively easy due to the documentation made available by Meta and by HuggingFace 1, the platform on which the model is hosted. It is also extremely convenient to create output on the fly, however, it still requires to have written text in the target language. No translation module is available yet for Mooré, therefore, one has to rely on a native speaker to provide the translations, which can take time, especially since the grammar can differ from region to region. The main downside of using AI is its resource consumption. Compared to the bot, it is more difficult and expensive in resources to host an AI model, so it is important to take into consideration when adding it to one's application, even if it is very useful.

12.7.2 Challenges Encountered

When developing the prototype, several challenges were encountered. The first, as mentioned above, is obtaining translations and audio recordings from native speakers (Stan et al. 2022). This often created delays because the next steps of development were based upon those recordings. Patchy internet sometimes meant that the person could not send the recordings when they wanted to. Moreover, writing translations can be a challenging task for a Mooré speaker because they are unsure of the grammar, if they are more used to writing in French for example. Delays were incurred when verifying the specific grammar with other people. There is also a notable lack of resources online on the Moor. language. The authors wanted to learn how to count and write numbers, and found online resources with contradictory spellings and sometimes rules. It was complex to create a generic code which could write out an integer.

Another challenge encountered was for handling the second use case for Artemisia. The structure of the bot was essentially the same, with a simpler data scheme. However, translating everything proved to be more difficult because no accurate enough AI model was found, therefore a lot of slot and filler audios had to be made. This needed some understanding of the Bambara language and where words fit. It is also required to understand the numbering system, which is admittedly simpler than Mooré. Overall, developing the bot in Bambara required non-trivial actions that took time, notably delays in getting audio recordings.

12.7.3 Discussion

Let us reflect on the usefulness of this research in its current context and how it can be generalised to broader domains. The project carried out in this thesis was just an example use case of how modern technologies can be leveraged in a relevant way for underserved rural communities of developing countries. This section delves into the potential of messaging bots in low-resource environments and how it could be improved in the future.

As seen in the developing and deploying phase, bots on messaging applications can be extremely cheap and easy to deploy due to the growing penetration rate of the Internet. Voice applications have been used for more than 10 years but they can be costly to implement as one needs to have a dedicated hotline. Therefore, voice systems for smartphones are a useful way to circumvent those costs and embrace current technologies while adapting it to its future users.

The project exhibited in this thesis is but one example of meaningful application in the domain of providing services to rural communities in West Africa. As seen from the context analysis, already four different use cases have emerged. And because of time constraints, the use cases were kept fairly simple by the first author. They were also designed to help a very specific community of women producing NTFPs and farmers selling Artemisia.

If the project is well validated by its intended users, it would be a good proof of concept for future research in implementing interactive voice systems for rural communities in developing countries. The advantages are many and could improve people's lives in many different contexts. Spreading ICTs to facilitate efficient and effective communication among community members could boost access to income (Bon et al. 2016), access to information about their profession (Morten et al. 2022; Norman and Draper 1986), about their education (The Django Software Foundation 2023; Dittoh et al. 2013), about their rights (Bon et al. 2020; Ortiz-Crespo et al. 2021) or even about their health (Madaio et al. 2020).

Those domains were already researched in different countries for voice-based applications, as recent papers show [ref ref]. A more mainstream option of creating and developing bots entirely online could boost the current research for more creative solutions.

AI particularly has shown great potential in helping reach people who speak local languages. More and more initiatives are taken to build models on languages less common (Marathe et al. 2015) and developing models which can be trained on minimal amounts of data (Gulaid and Vashistha 2013; Joshi et al. 2014).

The main downside of using AI is its complexity to run in low resource environments. Indeed, it AI models require considerable computing power that is more expensive and thus less accessible in for rural communities in developing countries. However, with technology advancing, it could be made more accessible in the future.

12.7.4 Future Work

To enhance the inclusivity and accessibility of the project, it is essential to support additional local languages spoken in Burkina Faso, such as Dyula, Bissa, and Fula. Incorporating these languages will ensure that the bot can cater to a broader audience, reflecting the linguistic diversity of the region. This extension will involve translating the user interface and voice messages into these languages, enabling more members to use the bot effectively.

By doing so, we can promote wider adoption and ensure that language barriers do not hinder the communication and cooperative efforts within rural communities. Implementing ASR functionalities in multiple languages can also significantly enhance the user experience of the Telegram bot. ASR technology can convert spoken language into text, allowing users to interact with the bot through voice commands, which is particularly useful in areas with low literacy rates. Developing ASR capabilities for languages such as Moor., Dyula, Bissa, and Fula will require collaboration with linguistic experts and leveraging machine learning models trained on these languages. This advancement will make the bot more intuitive and user-friendly, facilitating smoother and more efficient communication for all users, regardless of their literacy levels.

To further support cooperatives, the bot can be improved with additional functionalities that enable cooperative heads to manage their members more effectively. This could include features that allow cooperative heads to view the stock levels of each member, send automated messages to them requesting to update their current stock, and maintain an accurate, real-time inventory. By integrating these functionalities, cooperative administration can better oversee their operations, ensure timely updates from all members, and make informed decisions based on up-to-date information. These enhancements will streamline communication within the cooperative, reduce manual administrative tasks, and promote greater transparency and efficiency in stock management, ultimately contributing to the productivity and success of the cooperative selling model facilitated by the bot.

With the increase in internet and smartphone usage everywhere in the world, especially underserved communities, this project aimed at finding ways to leverage new technologies for novel purposes. Indeed, voice systems can easily be built and answer the needs of diverse communities in terms of literacy and digital literacy skills. Future works should focus on expanding the current system to ensure that it is valuable for its intended users. Moreover, the technologies touched upon in this chapter, namely building a bot on a messaging app and integrating AI models for speech processing, could be reused in many different use cases to further help more people. This has the capacity to increase the quality of life and access to information for many communities often forgotten about.

12.8   Conclusion

This study has shown how a digital application can be designed, built and deployed for users—in this case women cooperative in West Africa—in low resource environment, by *plugging in* the skills and technological knowledge of the digital

developer/researcher who is unfamiliar with the local context, into the local users', interest, knowledge and practices. First, the local context with all specific conditions is explored. With a better insight into the stakeholders and their operational goals, a digital solution (a so-called *bot*) can be effectively designed in close collaboration with the stakeholders, which plugs in into their interest, knowledge and practices, in this case the trade and knowledge about NTFP.

The approach is open-ended, flexible and iterative. The various stages can be carried out simultaneously or in a different order. As such, the project has implemented a prototype (bot) to facilitate and improve communication in rural communities of West African countries. This prototype serves as proof of concept for future research on communication in rural communities.

At the meta-level the approach followed in this research project has shown that it is possible to design and build digital solutions in co-creation with users in very low resource environments. The unfamiliarity with the local language, work processes and other features makes it essential and necessary to involve many local stakeholders. The digital solution is built according to the wish of local users. It has not been invented and brought in by some external expert or group.

This research also shows that the plug-in principle is an appropriate framework, also for digital development, as it integrates digital and local knowledge in a practical way, aware and respectful of what is already going on in the local context as invented by local users.

References

African Smallholder Farmers Group (2010) Africa's smallholder farmers: Approaches that work for viable livelihoods. Technical Report of the Food and Agriculture Organization of the United Nations. pp 1–28. https://www.fao.org/family-farming/detail/en/c/435252/. Accessed 19-10-2024

Ahenkan A, Boon E (2011) Improving nutrition and health through non-timber forest products in Ghana. J Health Popul Nutr 29(2):141–148

Bon A (2020) Intervention or collaboration?: redesigning information and communication technologies for development. PhD Thesis. Maastricht University, The Netherlands

Bon A, Akkermans JM, Gordijn J (2016) Developing ICT services in a low-resource development context. Complex Syst Inform Model Quart 9:84–109. https://doi.org/10.7250/csimq.2016-9.05. ISSN: 2255-9922 online. 6, 9, 18

Bon A, Gordijn J, Akkermans H (2020) E-service innovation in rural Africa through value co-creation. In: Disruptive technology, pp 859–877

Chengalur-Smith IS, Potnis D, Mishra G (2021) Developing voice-based information sharing services to bridge the information divide in marginalized communities: a study of farmers using IBM's spoken web in rural India. Int J Inf Manag 57:102283

Chou P (2018) The role of non-timber forest products in creating incentives for forest conservation: a case study of Phnom Prich Wildlife Sanctuary, Cambodia. Resources 7(3):41

de Boer V, De Leenheer P, Bon A, Gyan NB, van Aart C, Gueret C, Tuyp W, Boyera S, Allen M, Akkermans H, March R (2012) Distributed voice- and web- interfaced market information systems under rural conditions. In: Ralyt J, Franch X, Brinkkemper S, Wrycza S (eds) Advanced information systems engineering. Springer, Berlin, Heidelberg, pp 518–532

de Ridder S, van der Kooy F, Verpoorte R (2008) Artemisia annua as a self-reliant treatment for malaria in developing countries. J Ethnopharmacol 120(3):302–314

Dittoh F, van Aart C, de Boer V (2013) Voice-based marketing for agricultural products: a case study in rural northern Ghana. In: Proceedings of the sixth international conference on information and communications technologies and development: notes - volume 2. Association for Computing Machinery, New York, NY, USA, pp 21–24

Dittoh F, Akkermans H, de Boer V, Bon A, Tuyp W, Baart A (2022) Tibansim: information access for low-resource environments. In: Proceedings of sixth international congress on information and communication technology. Singapore, pp 675–683

Eizenga D (2023) Burkina Faso. Brill, Leiden, The Netherlands, pp 59–67

Gulaid M, Vashistha A (2013) Ila Dhageyso: an interactive voice forum to foster transparent governance in Somaliland. In: Proceedings of the sixth international conference on information and communications technologies and development: Notes - Volume 2. Association for Computing Machinery, pp 41–44

Heeks R (2017) Information and communication technology for development (ICT4D), 1st edn. Routledge, p 6

Joshi A, Rane M, Roy D, Emmadi N, Srinivasan P, Kumarasamy N, Pujari S, Solomon D, Rodrigues R, Saple DG, Sen K, Veldeman E, Rutten R (2014) Supporting treatment of people living with HIV/AIDS in resource limited settings with IVRs. In: Proceedings of the SIGCHI conference on human factors in computing systems. Association for Computing Machinery, pp 1595–1604

Karakara AA-W, Osabuohien ES (2022) Threshold effects of ICT access and usage in Burkinabe and Ghanaian households. Inf Technol Dev 28(3):511–531

Madaio MA, Yarzebinski E, Kamath V, Zinszer BD, Hannon-Cropp J, Tanoh F, Akpe YH, Seri AB, Jasińska KK, Ogan A (2020) Collective support and independent learning with voice-based literacy technology in rural communities. In: Proceedings of the 2020 CHI conference on human factors in computing systems. Association for Computing Machinery, pp 1–14

Marathe M, O'Neill J, Pain P, Thies W (2015) Revisiting CGNet Swara and its impact in rural India. In: Proceedings of the seventh international conference on information and communication technologies and development. Association for Computing Machinery

Mathieu N'dri L, Kakinaka M (2020) Financial inclusion, mobile money, and individual welfare: the case of Burkina Faso. Telecommun Policy 44(3):101926

Maua JO, Tsingalia HM, Cheboiwo J (2019) Economic value of non-timber forest products utilized by the households adjacent to the South Nandi forest reserve in Kenya. East Afr Agric For J 83(4):368–391

Morten BS, Haavik V, Iocchi A (2022) The end of stability—how Burkina Faso fell apart. Afr Secur 15(4):317–339

Norman DA, Draper SW (1986) User centered system design; new perspectives on human-computer interaction. L. Erlbaum Associates Inc., USA

Ortiz-Crespo JSB, van de Gevel J, Quirs CF, Daudi MGMH, van Etten J (2021) User-centred design of a digital advisory service: enhancing public agricultural extension for sustainable intensification in Tanzania. Int J Agric Sustain 19(5–6):566–582

Patel N, Chittamuru D, Jain A, Dave P, Parikh TS (2010) Avaaj Otalo: a field study of an interactive voice forum for small farmers in rural India. In: Proceedings of the SIGCHI conference on human factors in computing systems, CHI '10. Association for Computing Machinery, New York, NY, USA, pp 733–742

Prasad V, Shallam R, Sharma A, Varghese D, Mehta D (2021) A hybrid multi-modal system for conducting virtual workshops using interactive voice response and the WhatsApp business API. In: Extended abstracts of the 2021 CHI conference on human factors in computing systems. Association for Computing Machinery

Reij C, Waters-Bayer A (2014) Farmer innovation in Africa: a source of inspiration for agricultural development. Routledge

Stan GV, Baart A, Dittoh F, Akkermans H, Bon A (2022) A lightweight downscaled approach to automatic speech recognition for small indigenous languages. In: Proceedings of the 14th ACM web science conference 2022, WebSci '22. Association for Computing Machinery, New York, NY, USA, pp 451–458

The Django Software Foundation (2023) Django (Version 4.0) Software, available at https://www.djangoproject.com/

Wohlin C, Runeson P, Hst M, Ohlsson MC, Regnell B, Wessln A (2012) Experimentation in software engineering - an introduction. Kluwer Academic Publishers

Epilogue: The Future of Africa's Food Systems

Saa Dittoh, Anna Bon, and Hans Akkerman

According to Horowitz (1990), development interventions designed to enhance the habitat and improve the income and productivity of small-scale rural households in the African Sahel have often had exactly the opposite outcomes. That can be said for similar interventions in almost all parts of Africa. It has been so because of what the Chapters in this book have pointed out: the lack of adequate recognition given to indigenous and local knowledge.

The issue of integration of knowledge systems is not simply about "bottom-up" or participatory approaches. It is about giving farmers and community members the opportunity to take their own decisions about the future of their farming and livelihood activities. They should indeed be empowered to do so if one is aiming for the benefit of everyone. The role of governments, researchers, development workers (NGOs and CSOs), development partners and others, is to assist them, farmers and community members, to "better" their decisions, their technologies, their abilities and capabilities, and to remove the several unfreedoms that confront them as stated by Amartya Sen (Sen 1999). As documented throughout this book, however, the imposition from the outside, of systems, models and ("scientific") knowledge without regard of existing local, indigenous knowledge, experiences, contexts and interests *on-the-ground* has been generally faulty and continues to be so in Africa.

The future of food in Africa instead demands that the *"bettering processes"*—through agricultural training and extension delivery, research and innovations, infrastructural and input provision, development interventions, technological advancements etc.—are carried out differently with changed mindsets by relevant stakeholders, especially governments and their agents and development interventionists, including development partners.

The call for development initiatives to be done differently starts with recognizing the importance of what already exists, including indigenous knowledge and natural resources, and a resolve to improve, rather than replace, what nature (God) has conferred on different peoples, in terms of that knowledge and natural resources. That simple decision not to replace but to improve what exists has turned out to be the most difficult thing

S. Dittoh
University for Development Studies, Tamale, Ghana
e-mail: saaditt@gmail.com

A. Bon
Vrije Universiteit Amsterdam, Amsterdam, Noord-Holland, The Netherlands

H. Akkerman
University for Development Studies, Tamale, Ghana

AKMC Knowledge Management B.V., Koedijk, The Netherlands

© The Editor(s) (if applicable) and The Author(s) 2025
S. Dittoh et al. (eds.), *Integrating Indigenous and Scientific Knowledge for Sustainable Food Systems in Africa*, Sustainable Development Goals Series, https://doi.org/10.1007/978-3-031-85512-2

to do. Scientists have over time assumed that they know better than nature, and this *hybris* has proven to be wrong. Obvious evidence of that is climate change and its consequences.

Agricultural training and extension delivery in Africa has been about "what scientists think should be" and not about what exists and how what exists can be "bettered". Thus, it has been more about improving wheat, maize, rice, soybean, exotic vegetables and flowers, exotic poultry and commercial crops; rather than millet, sorghum, beans, cassava, yam, plantain/bananas, sweet potatoes, local vegetables, local poultry, guinea fowl and others. The call is for this to change, given the wide range of arguments in the book's Chapters with regard to resilience, sustainability, nutrition and biodiversity. Dittoh and Akuriba (2018) said the following about agricultural training in Africa:

> The training seems to be for "imaginary farms" and not for the African farms that exist. It is good to imagine "desirable" farms but if that is all and the reality on the ground is ignored, then the training becomes irrelevant.

Research and technological innovations also must be done quite differently so that they can align with what farmers do. Presently almost all agricultural research and technological innovations are about "commodities" rather than "systems" as explained in Chap. 2. Most agricultural research in Africa continues to be more of basic rather than applied research, which is what is required given the very limited resources and expertise that exist. Dietz et al. (2004) said the following about northern Ghanaian farmers:

> Many farmers combine a rather large variety of crops and crop varieties and various rounds of seeding, to spread risks and harvest moments; many farmers start with fast-maturing millets, followed by other millets and sorghum, and some maize and beans, often in the same fields, and with overlapping farm management cycles. During a drought they give up on more risky crops as far as drought stress is concerned and concentrate their efforts on crops and varieties which are more likely to succeed.

This process which works for the farmer cannot be researched by the methodologies commonly used. There have been attempts over time at research into improved indigenous soil practices (composting, crop residue management etc), intercropping, crop-livestock integration, agroforestry and others (Mutsaers et al. 1993; Berdjour et al. 2020; IITA 2021; MacLaren et al. 2021; Ongom et al. 2023). Most of them have, however, been the typical discipline-based and commodity-based research which does not properly reflect what the farmer does, and thus cannot improve upon his/her method of mixed cropping and mixed farming. Most current agriculture science research seems to be aiming at publications and not at the needs of the farmer. Support for research by development partners and others seem to be not aimed at the farmers, but for the benefit of multinational agribusinesses. The research is still largely conventional "green revolution" type, where the aim is to come up with improved technologies by the "experts" to be adopted by farmers (the non-experts). That is what the "*Plug-In Principle*" developed in this book has clearly shown to be problematic. Interdisciplinary and transdisciplinary research expertise must be developed to undertake research into agroecological and other sustainable agricultural methodologies.

There is a need that infrastructural and input provision to be different in Africa's future food systems. African infrastructural provision has been concentrated in urban areas. Roads, potable water and electricity provision have been biased against underserving rural areas where food is produced, preserved and processed. That must change within the context of the future of Africa's food systems, such that where food is produced has better infrastructure to ensure that everyone, especially the youth, will be encouraged to stay in the rural areas and become active participants, especially with respect to food production and processing.

Input provision should also follow a scenario different from the prevailing system. The current high cost of imported inputs should be deemphasized in favor of low external and locally available inputs which are far more cost-effective. The applicability of synthetic chemicals should be very limited and very precise to solve specific problems. The indigenous knowledge of valorizing crop residues and food wastes for the purpose of producing compost and other products should be improved upon. The provision of quality seed that is acceptable to farmers is another important change required in the future of African food systems. Seed and breed security is necessary for sustainable food and nutrition security. Farmers must have the liberty to decide what types of seed they want, and community members must be able to decide what food they will eat. That is, food sovereignty should be a necessary goal of the future of African food systems.

The greatest influence on agriculture and food interventions over several decades has been development interventions by development partners (DPs) including the World Bank and the African Development Bank. Their methods of intervention must change drastically. Most of the green revolution technologies which they continue to promote must give way to technologies that are nature-based and nature-inspired. Farmers in Africa have been expecting them to do so for a long time. It is unfortunate that it is taking so long for the necessary changes to be made.

Africa's past and present food problems have been complex due to many internal and external factors, including rapid population growth, lack of appropriate technologies, domestic and external self-seeking interests, particularly by the elite and multinational organizations, and lack of political will to address pertinent issues, as well as unexpected occurrences like conflicts and plagues (and now, climate change). The future of food in Africa can be even more complex and damaging if these complexities are not addressed

holistically, comprehensively and concurrently. Simply moving from green revolution technologies to agroecology and other sustainable technologies will not necessarily solve the problems that plague the continent. In the same vein, social, economic and self-seeking interests of the elite will not change overnight just because there is a change from green revolution technologies to "sustainable" technologies. There is a pressing need for targeted sensitization and advocacy to change mindsets and behaviours.

A future African food system must be resilient, sustainable and inclusive. The constituents of such a system may be summarized as follows:

- Ability to withstand environmental, conflict and disease shocks (resilience);
- Climate smart;
- Provides adequate planetary and ecosystem services health;
- Nutrition-sensitive, ensuring healthy diets for all;
- Gender-sensitive;
- Able to produce adequate, desirable and safe food in short, medium and long terms;
- Have profitable, socially-acceptable and environmentally-friendly value chains;
- Ensures healthy and sustainable consumption patterns; and
- Promotes equitable livelihoods (inclusiveness).

The possibility of a particular food system having all of the above desired attributes in a short term is not the issue. What is relevant for the future food system is to be moving towards these attributes and ambitions over time. As stated by Staaz and Dembele (2008), "Sub-Saharan Africa will need to develop a series of differentiated agricultural revolutions suited to its varied ecological niches and market opportunities".

References

Berdjour A, Dugje IY, Dzomeku IK, Rahman NA (2020) Maize–soybean intercropping effect on yield productivity, weed control and diversity in northern Ghana. Weed Biol Manag 20:69–81. https://doi.org/10.1111/wbm.12198

Dietz AJ, Millar M, Dittoh S, Obeng F, Ofori-Sarpong E (2004) Climate and livelihood change in North East Ghana. In: Dietz AJ, Ruben R, Verhagen A (eds) The impact of climate change on drylands, with a focus on West Africa. Kluwer Academic Publishers, Dordrecht/Boston/London

Dittoh S, Akuriba MA (2018) Africa's looming food and nutrition insecurity – A call for action. Ghana J Agric Econ Agribus 1(1):148–170

Horowitz MM (1990) Donors and deserts: the political ecology of destructive development in the sahel. Afr Environ VII(25-26-27-28):185–210. ENDA Third World, Dakar

IITA (2021) IITA News. No. 2598, 19–23 July

MacLaren C, Waswa W, Mead A, Claessens L, Vanlauwe B, Storkey J (2021) Optimising intercrops for western Kenya. Aspects Appl Biol 146:351–355. (Intercropping for sustainability: research developments and their application)

Mutsaers HJW, Ezumah HC, Osiru DSO (1993) Cassava-based intercropping: a review. Field Crops Res 34:431–457. International Institute of Tropical Agriculture, Ibadan, Nigeria

Ongom PO, Fatokun C, Togola A, Mohammed SB, Ishaya DJ, Bala G, Popoola B, Mansur A, Tukur S, Ibikunle M et al (2023) Exploiting the genetic potential of cowpea in an intercropping complex. Agronomy 13:1594. https://doi.org/10.3390/agronomy13061594

Sen A (1999) Development as freedom. OUP

Index